Lecture Notes in Chemistry

Edited by G. Berthier M. J. S. Dewar H. Fischer
K. Fukui G. G. Hall H. Hartmann H. H. Jaffé J. Jortner
W. Kutzelnigg K. Ruedenberg E. Scrocco

25

Paolo Arrighini

Intermolecular Forces and Their Evaluation by Perturbation Theory

Springer-Verlag
Berlin Heidelberg New York 1981

Author

Paolo Arrighini
Instituto di Chimica Fisica, Università di Pisa
Via Risorgimento, 35, I-56100 Pisa

ISBN-13:978-3-540-10866-5 e-ISBN-13:978-3-642-93182-6
DOI: 10.1007/978-3-642-93182-6

Softcover reprint of the hardcover 1st edition 1981

2152/3140-543210

To my wife Anna Maria, my daughter Silvia,
and my parents Bruna and Paolino.

PREFACE

The aim of these notes is to offer a modern picture of the perturbative approach to the calculation of intermolecular forces. The point of view taken is that a perturbative series truncated at a low order can provide a valuable way for evaluating interaction energies, especially if one limits oneself to the case of intermediate- and long-range distances between the interacting partners.

Although the situation corresponding to short distances is essentially left out from our presentation, the problems which are within the range of the theory form a vast and important class: a large variety of phenomena of matter, in fact, depends on the existence of interactions among atoms or molecules, which over a substantial range of distances should be classified as weak in comparison to the interactions occurring inside atoms or molecules.

We are aware of the omission of some topics, which in principle could have been included in our review. For instance, a very scarce attention has been paid to the analysis of problems involving interacting partners in degenerate states, which is of particular relevance in the case of interactions between excited atoms (only a rather quick presentation of the formal apparatus of degenerate perturbation theory is included in Chap.III). Interactions involving the simultaneous presence of more than two atoms (or molecules) have not been considered, with the consequent non-necessity of considering nonadditive effects which characterize the general N-body problem. In the same way, no attention has been paid to the so-called retardation effects, which play a role when the interacting atoms are very far from each other, due to the finite velocity of transmission of the electromagnetic interaction.

Even though we have stressed a perturbative approach, parts of these notes are hoped to be of value by themselves in the general context to the problems posed by the calculation of intermolecular forces. For example, the lengthy Chap.VII can be regarded as a fairly detailed source of material usable for evaluating second-order induction and dispersion energy contributions to the interaction energy between two atoms or molecules; dispersion energy contributions obtained in this way could then usefully be coupled to variational SCF calculations, which often provide reasonably reliable estimates for the remaining part of the interaction energy.

It would have been very difficult to put these notes in the form of a book without the help of Dr. C.Guidotti, who was available every moment for discussions and aid of any kind, of Mr. O.Cosci and Mrs. S.Grassini-Poli, who spent many hours at the keys of the typewriter: to all of them, and to Mr. P.Palla, who drew the figures for me impossible, I express my deep gratitude. A particular appreciation is due to Prof. E.Scrocco for his initial suggestion to write down these notes and to Springer-Verlag for patiently waiting for my never-ending manuscript. Finally I wish to thank my family for serenely accepting my engagement for a long period indeed, even though it signified a renunciation to spare time which we could have spent together.

CONTENTS

I. INTRODUCTION

The extremely important role of intermolecular forces in determining equilibrium and non-equilibrium properties and phenomena of matter is by now a well established point. In spite of this recognition and the awareness that quantum mechanics is the natural framework to be used for a proper understanding of intermolecular interactions, a fully rigorous approach for these is not yet well established, especially if one pursues the ambitious program of dealing with the whole range of separations between the interacting partners.

In its essence, the problem we are faced with is a many-electron one, in the presence of nuclei at a fixed geometry (Born-Oppenheimer approximation) and the program is solving the associate Schrödinger equation for the allowed energy levels $E_0(\{\underline{R}\})$, $E_1(\{\underline{R}\}), \ldots$, where $\{\underline{R}\}$ denotes all the independent nuclear coordinates. If we imagine this procedure to be repeated for innumerable nuclear configurations $\{\underline{R}\}$, the quantities $E_n(\{\underline{R}\})$ as a function of $\{\underline{R}\}$ represent hypersurfaces, that "pass" through the values $E_n(\{\underline{R}_\infty\})$, corresponding to the energies of the involved partners as completely far apart (often referred to as "monomers" or subsystems). One refers frequently to the quantity $E_n(\{\underline{R}\})$ as the energy of the "complex" or "supermolecule" and $\Delta E_n = E_n(\{\underline{R}\}) - E_n(\{\underline{R}_\infty\})$ is one of the searched interaction energies.

Clearly the task is, in general, formidable, even though we limit ourselves to the ground state, and one can only hope for sensible, approximate solutions to the Schrödinger equation, essentially to be got by variational or perturbative techniques.

Although the variational methods, by which the problem of chemical interactions has been approached traditionally, have successfully been applied also in the present context and therefore hold a rightful place among the methodological approaches to the problem of intermolecular interactions, they will be left out from the present review, which is entirely devoted to a presentation of the perturbative theory.

It is right to say that our choice does not imply in any way a recognition of superiority of the perturbative approach with respect to the variational one, but reflects the desire of making better known a subject, which along with intriguing features has surely various positive elements. As we shall see, in fact, in due time, the ordinary Rayleigh-Schrödinger perturbation theory does not work when applied to the problem at issue, because generally unable to connect the unperturbed ground state of the non-interacting partners with the ground state of the "complex", for any configuration, so that some new perturbation theory must be devised, which overcomes this serious trouble.

A second difficulty is that there is not a unique theory of this kind, and the various theories are surely not equivalent if truncated after some finite order: a practical (i.e. limited to the first few orders) utilization of the perturbative approach involves, therefore, the ascertainment of an optimal theory (if there exists).

There are certainly some points which make a perturbation theory of molecular interaction energies very gratifying:

a) The interaction energy ΔE, usually a rather small quantity, is obtained directly, unlike the variational approach, where the same quantity arises, through a critical subtraction, from two nearly equal energies $E(\{\underline{R}\})$ and $E(\{\underline{R}_\infty\})$ of the "complex" and the non-interacting partners, respectively.

b) The interaction energy comes out decomposed in a very natural way into various physically meaningful contributions, which behave in a different manner as the relative separations between the interacting partners vary. This feature is quite generally important, but becomes crucial if one has in mind an extension of the approach to situations (out of our present reach) where relatively large molecules are involved.

It does not seem bold to assert that a (low-order) perturbation theory of intermolecular interactions is in general of noticeable conceptual value and of practical value in the regions corresponding to intermediate and long-range separations between the partners. As far as the latter point is concerned, however, more calculations are surely required, because a firmer confidence can be gained only through a desirable uniform reliability of results: from this standpoint it is also clear that the parallel development of accurate variational calculations is essential as a tool for providing us with credible standards of comparison.

From the preceding remarks it should be manifest how these notes are not a comprehensive review about intermolecular forces. Excellent books and articles on the subject are already available, where more general goals are pursued and the pedagogical intent more accentuated; in particular, we recommend warmly the following ones:

[1] Hirschfelder, J.O., Curtiss, C.F. and Bird, R.B., Molecular Theory of Gases and Liquids, Wiley, N.Y., 1964.

[2] Hirschfelder, J.O. (ed.), Intermolecular Forces, Adv. Chem. Phys. XII, Interscience, 1967.

[3] Buckingham, A.D. and Utting, B.D. in Ann. Rev. of Phys. Chem. 21, 287 (1970).

[4] Margenau, H. and Kestner, N.R., Theory of Intermolecular Forces, 2-nd Ed., Pergamon Press, London, 1971.

[5] Certain, P.R. and Bruch, L.W. in Intern. Rev. of Science, Phys. Chem. Series 1, Vol. 1, Butterworths, London, 1972.

[6] Pullman, B. (ed.), Intermolecular Interactions: from Diatomics to Biopolymers, Wiley, N.Y., 1978.

All these references will be listed in the bibliography by using the same numbers as here.

II. SYMMETRY: AN EXCURSION THROUGH ITS FORMAL APPARATUS

II.1. Constants of Motion and Projectors.

Symmetry and invariance properties permeate quantum mechanics in a fundamental way. The relevant symmetry, that of the Hamiltonian operator $\hat{H}$, is to be traced back, as well known, to the existence of one or more operators commuting with $\hat{H}$ itself.

Any operator $\hat{\Lambda}$ which commutes with the Hamiltonian $\hat{H}$ of a given system, i.e. such that

$$[\hat{H},\hat{\Lambda}] = 0 \ ,$$

is said a constant of motion of the system*. If moreover $\hat{\Lambda}$ commutes with its adjoint, i.e.

$$[\hat{\Lambda},\hat{\Lambda}^{\dagger}] = 0 \ ,$$

it will be called a normal constant of motion (self-adjoint as well as unitary constants of motion therefore belong to this class) [7,8].

A set of constants of motion $\{\hat{\Lambda}_i\}$ is a compatible set if $[\hat{\Lambda}_i,\hat{\Lambda}_j]=0$, $\forall i,j$. A complete set of compatible constants of motion constitutes the standard way of labeling the degenerate states associated with a given energy level (the eigenvalues of the set of operators $\{\hat{\Lambda}_i\}$).

In many cases the symmetry of the Hamiltonian operator corresponds to constants of motion which are not compatible, i.e. non-commutative. It is not difficult to show that the set $G \equiv \{\hat{g}_i\}$ formed by all constants of motion which are also unitary is a group [9,10], the so-called unitary group of the Hamiltonian. In the (special) case the group G is abelian, we fall back into the case of compatible constants of motion. In general, however, the group G is non-commutative; in such a case, the quantum states cannot be classified by means of eigenvalues, but by the irreducible representations Γ^{α} of the group and its subgroups.

Let λ_ℓ denote the generic eigenvalue of the constant of motion $\hat{\Lambda}$. The following quantity

$$\hat{Q}_k = \prod_{\ell \neq k} \frac{\hat{\Lambda} - \lambda_\ell \hat{1}}{\lambda_k - \lambda_\ell} \qquad \text{(II.1.1)}$$

*We shall assume that $\hat{\Lambda}$ does not depend explicitly on time.

($\hat{1}$ denotes the identity operator) is then a <u>projection</u> <u>operator</u>, which "extracts" from a general state $|\Phi\rangle$ of the Hilbert space its component along the k-th eigenvector $|\lambda_k\rangle$ to $\hat{\Lambda}$.

The set $\{\hat{Q}_k\}$ of projection operators associated with the spectrum of eigenvalues $\{\lambda_k\}$ to $\hat{\Lambda}$ satisfies the following basic relations [10]

$$
\begin{array}{lll}
\text{i)} & \hat{\Lambda}\hat{Q}_k = \lambda_k \hat{Q}_k & (\hat{Q}_k \text{ is an eigenoperator to } \hat{\Lambda} \text{ with eigenvalue } \lambda_k) \\
\text{ii)} & \hat{H}\hat{Q}_k = \hat{Q}_k \hat{H} & (\hat{Q}_k \text{ commutes with } \hat{H}\text{, i.e. it is } \underline{\text{itself}} \text{ a constant of motion}) \\
\text{iii)} & \hat{Q}_k^2 = \hat{Q}_k & (\hat{Q}_k \text{ is idempotent}) \\
\text{iv)} & \hat{Q}_k^\dagger = \hat{Q}_k & (\hat{Q}_k \text{ is self-adjoint}) \\
\text{v)} & \hat{Q}_k \hat{Q}_\ell = 0 \quad (\text{for } k \neq \ell) & (\text{the projectors in the set } \{\hat{Q}_j\} \text{ are mutually exclusive}) \\
\text{vi)} & \sum_k \hat{Q}_k = \hat{1} & (\text{resolution of the identity})
\end{array}
\qquad \text{(II.1.2)}
$$

The last relation plays a particularly important role, because it allows an arbitrary state $|\Phi\rangle$ to be analyzed in terms of eigenvectors to $\hat{\Lambda}$; in fact,

$$|\Phi\rangle \equiv \hat{1}\,|\Phi\rangle = (\sum_k \hat{Q}_k)\,|\Phi\rangle = \sum_k \hat{Q}_k\,|\Phi\rangle = \sum_k |\Phi_k\rangle \ ,$$

and $|\Phi_k\rangle \equiv \hat{Q}_k|\Phi\rangle$ satisfies $\hat{\Lambda}|\Phi_k\rangle = \lambda_k|\Phi_k\rangle$.

A fundamental service is done by the set of projection operators $\{\hat{Q}_k\}$. As a matter of fact, if $|\Phi_k\rangle$ and $|\Psi_\ell\rangle$ are vectors belonging to the eigenvalues λ_k and λ_ℓ, so that $\hat{Q}_k|\Phi_k\rangle = |\Phi_k\rangle$, $\hat{Q}_\ell|\Psi_\ell\rangle = |\Psi_\ell\rangle$, we obtain, for $k \neq \ell$,

$$
\begin{aligned}
\langle\Phi_k|\Psi_\ell\rangle &= \langle\Phi_k|\hat{Q}_k^\dagger\ \hat{Q}_\ell|\Psi_\ell\rangle = 0 \\
\langle\Phi_k|\hat{H}|\Psi_\ell\rangle &= \langle\Phi_k|\hat{Q}_k^\dagger\ \hat{H}\hat{Q}_\ell|\Psi_\ell\rangle = 0
\end{aligned}
\qquad \text{(II.1.3)}
$$

an immediate consequence of the relations (II.1.2). Eqs. (II.1.3) show that the Hilbert space is splitted by the set of projection operators

$\{\hat{Q}_k\}$ into subspaces (eigenspaces to $\hat{\Lambda}$, with eigenvalues $\{\lambda_k\}$) orthogonal to each other and non-interacting with respect to $\hat{H}$.

If the Hamiltonian $\hat{H}$ admits several compatible constants of motion $\{\hat{\Lambda}_i\}$, the preceding statements can directly be generalized by introducing a sequence of associated projectors $\{\hat{Q}_k(\hat{\Lambda}_1)\}$, $\{\hat{Q}_\ell(\hat{\Lambda}_2)\}$,.....

II.2. A Review of Group Theory Concepts.

In order to extend our considerations to the case of not compatible constants of motion, it is profitable to make use of some tools taken from group theory. Since books and comprehensive articles on group theory and its applications are numerous ([9,11-15] are only a very reduced list), here we shall confine ourselves, in armony with the spirit of this chapter, to a concise review of some concepts and results, as descend from a rather formal approach, based on the group algebra.

With any group $G \equiv \{\hat{g}_1 .....\hat{g}_n\}$ of operators, we may associate an algebra $A(G)$, called the (Frobenius) group algebra, defined as the linear space spanned by the n group elements $\hat{g}_1,...,\hat{g}_n$ regarded as basis vectors. An element of $A(G)$ has, therefore, the form

$$\hat{a} = a_1\hat{g}_1 + a_2\hat{g}_2 + ... + a_n\hat{g}_n = \sum_{i=1}^{n} a_i\hat{g}_i \quad ,$$

a_i being an arbitrary complex number. The product $\hat{a}\hat{b}$ of two elements of $A(G)$ is so defined:

$$\hat{a}\hat{b} = (\sum_i a_i\hat{g}_i)(\sum_j b_j\hat{g}_j) = \sum_{ij} a_i b_j (\hat{g}_i\hat{g}_j) \quad ,$$

which is still an element of the group algebra, because $(\hat{g}_i\hat{g}_j)$ belongs to G.

The group elements $\{\hat{g}_1,...,\hat{g}_n\}$ can be regarded in a twofold way: on the one hand, they are basis vectors for the group algebra; on the other, they are operators acting in the linear space $A(G)$, their effect being defined by the group multiplication table.

A linear subspace of the algebra $A(G)$ which is invariant (or stable) under the operators of the group (when operating on the left) is called a <u>(left) ideal</u>. The group algebra A(G) is clearly a linear space invariant under the group operators $\hat{g}_i$. The effect of $G = \{\hat{g}_1 \ldots . \hat{g}_n\}$ in the algebra A(G) is, therefore, represented by a set of n-dimensional matrices $\Gamma^R(\hat{g}_1), \ldots, \Gamma^R(\hat{g}_n)$, which constitute the so-called <u>regular representation</u> of the group. If now $\{\hat{g}_1, \ldots, \hat{g}_n\}$ are regarded as basis vectors spanning the group algebra, then $[\Gamma^R(\hat{g}_i)]_{jk}$ gives the "component" of $\hat{g}_i\hat{g}_k$ along the basis vector $\hat{g}_j$, so that $[\Gamma^R(\hat{g}_i)]_{jk} = 0$ if $\hat{g}_i\hat{g}_k \neq \hat{g}_j$ and $[\Gamma^R(\hat{g}_i)]_{jk} = 1$ if $\hat{g}_i\hat{g}_k = \hat{g}_j$. The elements of the matrix $\Gamma^R(\hat{g}_i)$ are, therefore, either zero or one and each row as well as each column of such matrix has only a single non-zero element.

If we pass to the characters $\chi^R(\hat{g}_i)$, it is not difficult to show that $\chi^R(\hat{1}) = n$, $\chi^R(\hat{g}_i) = 0$ $(\hat{g}_i \neq \hat{g}_1)$, where $\hat{g}_1 \equiv \hat{1}$ denotes the identity operator. It is immediate to decide that for $n > 1$ the invariant linear space A(G) and the associated matrix representation Γ^R are reducible. In order to determine how many times Γ^α is contained in the regular representation Γ^R, from the basic character formula $a_\alpha = n^{-1}\sum_{j=1}^{n}[\chi^\alpha(\hat{g}_j)]^*\chi^R(\hat{g}_j)$ we get $a_\alpha = n^{-1}\chi^\alpha(\hat{g}_1)\chi^R(\hat{g}_1) = \chi^\alpha(\hat{g}_1) = \lambda_\alpha$, λ_α being the dimension of the α-th irreducible representation of G. Thus the regular representation Γ^R contains each irreducible representation Γ^α of the group G as many times as its dimension. From such result it easily follows that $\sum_{\alpha=1}^{n_\Gamma}\lambda_\alpha^2 = n$, n_Γ being the number of irreducible representations of the group.

Since the regular representation (which provides the effect of the group operators in the algebra A(G)) is reducible, the linear space A(G) is also decomposable. The decomposition of A(G) generates λ_α independent irreducible (left) ideals $\mathcal{L}_1^\alpha, \mathcal{L}_2^\alpha, \ldots, \mathcal{L}_{\lambda_\alpha}^\alpha$ corresponding to the irreducible representation Γ^α. Thus the overall splitting of the algebra is expressed by the sum

$$A(G) = \sum_{\alpha=1}^{n_\Gamma}(\mathcal{L}_1^\alpha + \mathcal{L}_2^\alpha + \ldots + \mathcal{L}_{\lambda_\alpha}^\alpha),$$

each (left) ideal $\mathcal{L}_s^\alpha$ being spanned by a set of elements of A(G) which transform as partners of the irreducible representation Γ^α.

In order to find a set of elements of the group algebra which span a (left) ideal $\mathcal{L}_s^\alpha$ belonging to Γ^α, let us consider the following λ_α

elements (operators) of A(G)

$$\hat{\rho}^{\alpha}_{rs} = \frac{\lambda_\alpha}{n} \sum_{j=1}^{n} [\Gamma^{\alpha}(\hat{g}_j^{-1})]_{sr}\, \hat{g}_j \quad \left\{ \begin{matrix} r = 1,2,\ldots,\lambda_\alpha \\ s \text{ fixed} \end{matrix} \right\} \tag{II.2.1}$$

If we multiply this quantity (on the left) by the generic element $\hat{g}_k$ of G, by using the fact that $\hat{g}_k\hat{g}_j$ is an element $\hat{g}_i$ of G and $\Gamma^{\alpha}(\hat{g}_i^{-1}\hat{g}_k) = \Gamma^{\alpha}(\hat{g}_i^{-1})\,\Gamma^{\alpha}(\hat{g}_k)$, one finds

$$\hat{g}_k\hat{\rho}^{\alpha}_{rs} = \frac{\lambda_\alpha}{n} \sum_{i=1}^{n} \sum_{q=1}^{\lambda_\alpha} [\Gamma^{\alpha}(\hat{g}_i^{-1})]_{sq} [\Gamma^{\alpha}(\hat{g}_k)]_{qr}\hat{g}_i = \sum_{q=1}^{\lambda_\alpha} [\Gamma^{\alpha}(\hat{g}_k)]_{qr}\hat{\rho}^{\alpha}_{qs} \tag{II.2.2}$$

We have thus verified that the operators $\hat{\rho}^{\alpha}_{1s}, \hat{\rho}^{\alpha}_{2s}, \ldots, \hat{\rho}^{\alpha}_{\lambda_\alpha s}$ transform under G in accordance with the irriducible representation Γ^{α} of the group and the subspace $\mathcal{L}^{\alpha}_{s}$ spanned by them is an irreducible (left) ideal (belonging to Γ^{α}). As the index s ranges over the values $1,2,\ldots,\lambda_\alpha$, we can construct λ_α basis sets spanning the ideals $\mathcal{L}^{\alpha}_{1},\ldots\mathcal{L}^{\alpha}_{\lambda_\alpha}$, i.e. λ^2_α vectors associated with Γ^{α} and, therefore, a totality of $\sum_{\alpha=1}^{n_\Gamma} \lambda^2_\alpha = n$ vectors, as many as the dimension of the group algebra; the elements $\hat{\rho}_{rs}$ $(r,s = 1,2,\ldots,\lambda_\alpha;\ \alpha = 1,\ldots n_\Gamma)$ consequently span the entire algebra A(G).

The group $G \equiv \{\hat{g}_1,\ldots,\hat{g}_n\}$ will hereafter be assumed to be unitary: as already stated, in fact, the usual (symmetry) transformations which leave the Hamiltonian operator invariant can be associated with unitary operators. Then, $\hat{g}_j^{-1} = \hat{g}_j^{\dagger}$ and $[\Gamma^{\alpha}(\hat{g}_j^{-1})]_{sr} = [\Gamma^{\alpha}(\hat{g}_j)]^*_{rs}$, so that the quantities $\hat{\rho}^{\alpha}_{rs}$ can be cast in the form

$$\hat{\rho}^{\alpha}_{rs} = \frac{\lambda_\alpha}{n} \sum_{j=1}^{n} [\Gamma^{\alpha}(\hat{g}_j)]^*_{rs}\, \hat{g}_j \tag{II.2.3}$$

from which one easily verifies that $\hat{\rho}^{\alpha\dagger}_{rs} = \hat{\rho}^{\alpha}_{sr}$.

We shall now derive a very important relation satisfied by any pair of elements $\hat{\rho}^{\alpha}_{rs}$, $\hat{\rho}^{\beta}_{pq}$ which respectively transform in accordance with the irreducible representations Γ^{α}, Γ^{β} of the group G. To this aim, let us consider the operator product $\hat{\rho}^{\alpha}_{rs}\,\hat{\rho}^{\beta}_{pq}$; from the preceding equations we get

$$\hat{\rho}^{\alpha}_{rs}\,\hat{\rho}^{\beta}_{pq} = \frac{\lambda_\alpha}{n} \sum_{j=1} [\Gamma^{\alpha}(\hat{g}_j)]^*_{rs}\hat{g}_j\hat{\rho}^{\beta}_{pq} = \frac{\lambda_\alpha}{n} \sum_{j=1} \sum_{t=1}^{\lambda_\beta} [\Gamma^{\alpha}(\hat{g}_j)]^*_{rs} [\Gamma^{\beta}(\hat{g}_j)]_{tp}\, \hat{\rho}^{\beta}_{tq}$$

which, in virtue of the "strong" orthogonality relation for the matrix elements of irriducible representations [9,11],

$$\sum_{j=1}^{n} [\Gamma^{\alpha}(\hat{g}_j)]^*_{rs} [\Gamma^{\beta}(\hat{g}_j)]_{tp} = \frac{n}{\lambda_\alpha} \delta_{\alpha\beta} \delta_{rt} \delta_{ps}$$

can be rewritten as

$$\hat{\rho}^{\alpha}_{rs} \hat{\rho}^{\beta}_{pq} = \delta_{\alpha\beta} \delta_{sp} \hat{\rho}^{\beta}_{rq} \qquad \text{(II.2.4)}$$

Eq. (II.2.4) is a very important result for application to quantum mechanics.

It is convenient to regard the quantities $\hat{\rho}^{\alpha}_{rs}$ (corresponding to a given α) as matrix elements, the first index r referring to the row and the second one s to the column of the matrix. For the diagonal elements, from eq. (II.2.4) one easily obtains

$$\hat{\rho}^{\alpha}_{ss} \hat{\rho}^{\alpha}_{qq} = \delta_{sq} \hat{\rho}^{\alpha}_{ss} \qquad \text{(II.2.4')}$$

Thus, the λ_α elements $\{\hat{\rho}^{\alpha}_{ss}\}$ constitute a set of self-adjoint, mutually annulling, <u>projection operators</u>, having the important characteristics of generating irreducible (left) ideals in A(G).

In addition to the projection operators $\hat{\rho}^{\alpha}_{ss}$, whose construction requires the detailed knowledge of the representation matrices $\{\Gamma^{\alpha}(\hat{g}_j)\}$, other "weaker" projection operators,

$$\hat{\rho}^{\alpha} \equiv \sum_{s=1}^{\lambda_\alpha} \hat{\rho}^{\alpha}_{ss} = \frac{\lambda_\alpha}{n} \sum_{s=1}^{\lambda_\alpha} \sum_{j=1}^{n} [\Gamma^{\alpha}(\hat{g}_j)]^*_{ss} \hat{g}_j = \frac{\lambda_\alpha}{n} \sum_{j=1}^{n} \chi^{\alpha}(\hat{g}_j)^* \hat{g}_j \qquad \text{(II.2.5)}$$

are easily built up from the characters $\chi^{\alpha}(\hat{g}_j)$ only. $\hat{\rho}^{\alpha}$ is generally referred to as a <u>character projection operator</u>.

A fundamental property of the projection operators is that

$$\sum_{\alpha=1}^{n_\Gamma} \sum_{s=1}^{\lambda_\alpha} \hat{\rho}^{\alpha}_{ss} = \sum_{\alpha=1}^{n_\Gamma} \hat{\rho}^{\alpha} = \hat{1} \ , \qquad \text{(II.2.6)}$$

i.e. they provide a resolution of the identity [9].

In order to better appreciate the role of the projection operators, let us consider their action on some arbitrary state $|\Psi\rangle$ of the Hilbert space. From eq. (II.2.6) one obtains the following resolution into components

$$|\Psi\rangle \equiv \hat{1}|\Psi\rangle = \sum_{\alpha=1}^{n_\Gamma} \sum_{s=1}^{\lambda_\alpha} \hat{\rho}^\alpha_{ss}|\Psi\rangle \equiv \sum_{\alpha=1}^{n_\Gamma} \sum_{s=1}^{\lambda_\alpha} |\Psi^\alpha_{ss}\rangle \quad ,$$

with transformation properties under the group G

$$\hat{g}_k|\Psi^\alpha_{ss}\rangle = \hat{g}_k\ \hat{\rho}^\alpha_{ss}|\Psi\rangle = \sum_{q=1}^{\lambda_\alpha} [\Gamma^\alpha(\hat{g}_k)]_{qs}|\Psi^\alpha_{qs}\rangle \quad .$$

The projector $\hat{\rho}^\alpha_{ss}$ therefore generates a component $|\Psi^\alpha_{ss}\rangle$ which under the group of operators G transforms according to the s-th column of the irreducible representation Γ^α of G.

We are exclusively interested in the case where the group G consists of unitary operators commuting with the Hamiltonian operator $\hat{H}$, so that $[\hat{g}_i,\hat{H}] = 0$. Then also $[\hat{\rho}^\alpha_{ss},\hat{H}] = [\hat{\rho}^\alpha,\hat{H}] = 0$. Therefore, from eq. (II.2.4) and $\hat{\rho}^{\alpha\dagger}_{ss} = \hat{\rho}^\alpha_{ss}$, we get

$$\langle\Psi^\alpha_{ss}|\Psi^\beta_{rr}\rangle = \langle\Psi|\hat{\rho}^\alpha_{ss}\ \hat{\rho}^\beta_{rr}|\Psi\rangle = \delta_{\alpha\beta}\ \delta_{rs}\ \langle\Psi|\hat{\rho}^\alpha_{ss}|\Psi\rangle$$

$$\langle\Psi^\alpha_{ss}|\hat{H}|\Psi^\beta_{rr}\rangle = \langle\Psi|\hat{\rho}^\alpha_{ss}\hat{H}\hat{\rho}^\beta_{rr}|\Psi\rangle = \delta_{\alpha\beta}\ \delta_{rs}\ \langle\Psi|\hat{H}\ \hat{\rho}^\alpha_{rr}|\Psi\rangle \ .$$

The Hilbert space is therefore decomposed by the projectors $\hat{\rho}^\alpha_{ss}$ corresponding to different irreducible representations of G into orthogonal subspaces, which are also non-interacting with respect to $\hat{H}$.

The importance of the previous results for simplifying the complexity of secular problems can hardly be emphasized.

II.3. A Look to the Symmetric Group.

This rather hard subject, which is covered in several books on group theory [12-15], can be tasted, for a first adequate approach, in the very pedagogical introduction by Coleman [16]. In view of that, we shall not dwell on this topic too much, limiting ourselves to recall some of the most important concepts.

The symmetric group S_k is the group whose elements are the k! possible permutation operations $\hat{P}_i$ on k objects. Thus, S_3 contains the six elements $\hat{P}_1 = \hat{1}$, $\hat{P}_2 = (12)$, $\hat{P}_3 = (13)$, $\hat{P}_4 = (23)$, $\hat{P}_5 = (123)$, $\hat{P}_6 = (132)$, where, for example, the transposition $(12) \equiv (12)(3) \equiv \begin{pmatrix} 1 & 2 & 3 \\ 2 & 1 & 3 \end{pmatrix}$.

A useful property of any permutation is that it can be expressed (in a non-unique way) as a product of transpositions. For instance, $(123) = (13)(12) = (23)(13)$, $(12345) = (15)(14)(13)(12) = (45)(15)(23)(13) = \ldots$, etc. A permutation is said even (odd) if it is expressed as a product of an even (odd) number of transpositions.

Concerning the irreducible representations of S_k, for $k > 1$ every symmetric group has two one-dimensional irreducible representations: Γ^s, the fully symmetric one, where each group element is associated with +1 and Γ^a, the fully antisymmetric representation, in which the even (odd) permutations are represented by +1 (-1). The group S_2 possesses only the two representations Γ^s and Γ^a, while for S_k $(k > 2)$ Γ^s and Γ^a are the only one-dimensional representations.

If Γ^α is a representation (dimension λ_α), then the direct product $\bar{\Gamma}^\alpha = \Gamma^\alpha \otimes \Gamma^a$ (also of dimension λ_α) is another representation of S_k. Clearly, for even permutations $\hat{P}_i^e$, $\bar{\Gamma}^\alpha(\hat{P}_i^e) = \Gamma^\alpha(\hat{P}_i^e)$, but for odd permutations $\hat{P}_i^o$, $\bar{\Gamma}^\alpha(\hat{P}_i^o) = -\Gamma^\alpha(\hat{P}_i^o)$. Γ^α and $\bar{\Gamma}^\alpha$ are called associate representations. A very important theorem concerning associate representations states that the direct product $\Gamma^\alpha \otimes \Gamma^\beta$ of two irreducible representations of S_k contains the fully antisymmetric one Γ^a once and only once if Γ^α and Γ^β are associate and not at all if Γ^α and Γ^β are not associate [15]. The proof of the theorem, rather simple, can be obtained by standard application of the character formula.

An extremely important way of labelling the (inequivalent) irreducible representations of S_k is offered by the so-called Young diagrams (or graphs), which are arrays of k boxes arranged in rows, η_1 in the first row, η_2 in the second one, etc., in such a way that no row has more boxes than the rows above. For instance, for S_2 we have the following Young diagrams:

analogously, for S_3:

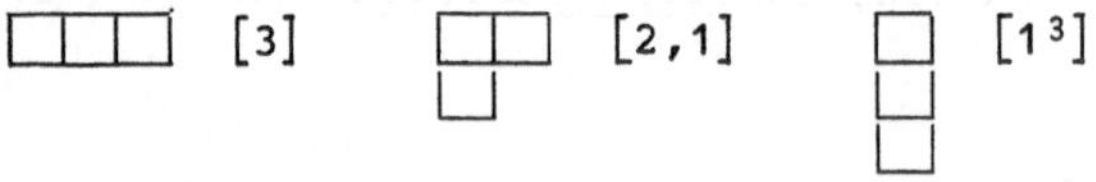

and for S_4:

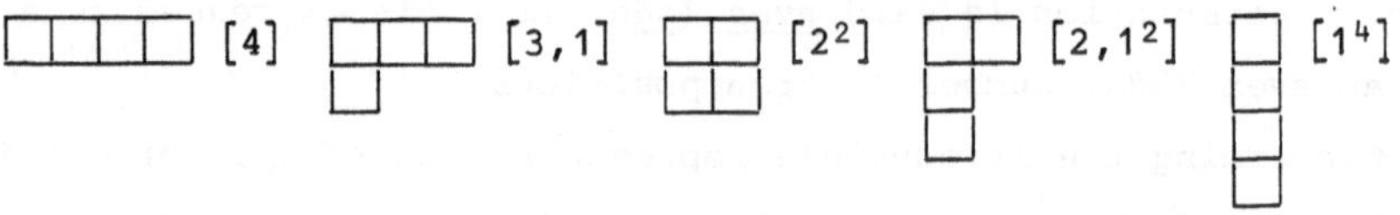

The notation which accompanies each graph should be obvious: $[2,1^2] \equiv [2,1,1]$ means two boxes in the first row, one in the second and one in the third one, and so on for the other diagrams. Therefore, the group S_3 admits three (inequivalent) irreducible representations, S_4 five, etc.

If we now insert numbers from one to k in each diagram for S_k, in such a way they increase from left to right in a row and from the top to the bottom in a column, we get the so-called (Young) standard tableaux. So, in correspondence with the Young graphs $[\tau]$ represented above, we have the following standard tableaux

S_2: | 1 | 2 |

    | 1 |
    | 2 |

S_3: | 1 | 2 | 3 |

    | 1 | 2 |
    | 3 |

    | 1 | 3 |
    | 2 |

    | 1 |
    | 2 |
    | 3 |

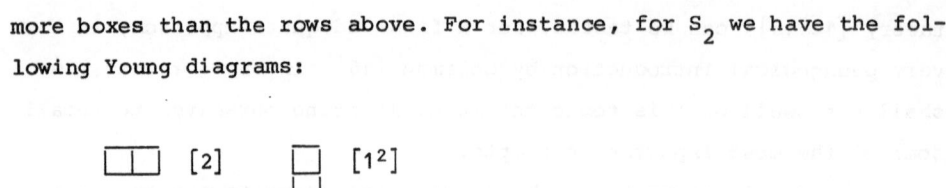

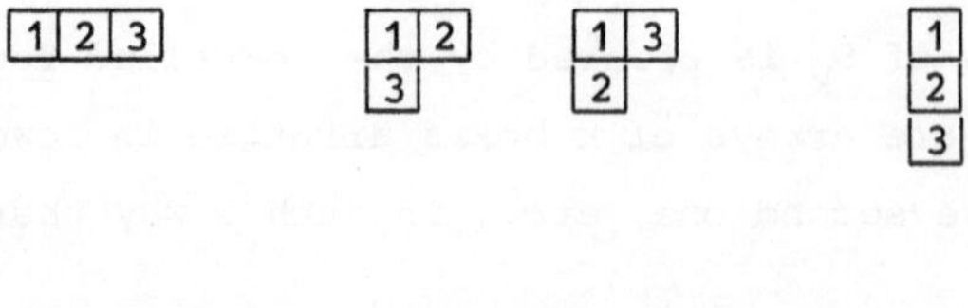

S_4: 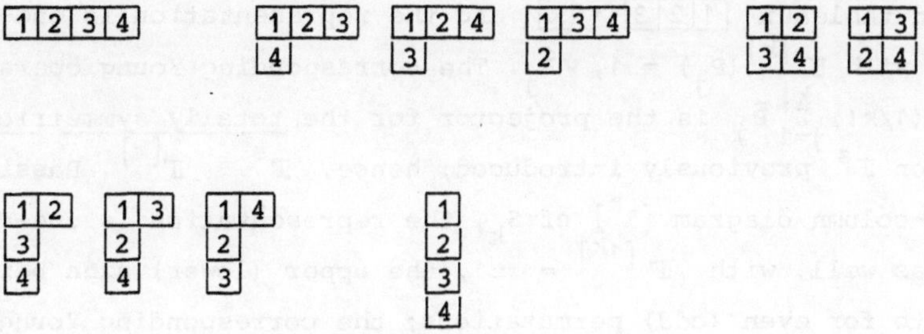

The number of standard tableaux corresponding to a given graph provides the dimensionality of the irreducible representation involved ; so, for instance, for S_4 the irreducible representation $[2^2]$ is two-dimensional, $[3,1]$ three-dimensional, etc. One verifies, in particular, that every group S_k $(k > 1)$ has two (and only two) one-dimensional irreducible representations, $[k] \equiv$ 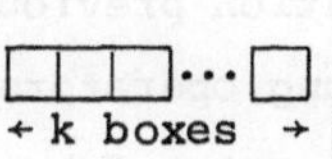 and $[1^k] \equiv$ , respectively.

It is not difficult to show that the <u>transposed diagram</u> $[\tilde{\tau}]$, obtained by interchanging the row and the columns of the diagram $[\tau]$, <u>corresponds to the associate irreducible representation</u>.

Let us now consider the symmetric group S_k and its algebra $A(S_k)$. The general results obtained in the previous section will keep their validity and we expect there exist operators which generate irreducible ideals of the algebra $A(S_k)$, in each of them the effect of the group operators being furnished by an irreducible representation of the group S_k. From eq. (II.2.3) and the fact that the representations for S_k can be assumed to be real [13], we obtain

$$\hat{\rho}^{[\tau]}_{rs} = \frac{\lambda_\tau}{k!} \sum_{j=1}^{k!} \Gamma^{[\tau]}_{rs}(\hat{P}_j)\hat{P}_j \qquad \text{(II.3.1)}$$

λ_τ being the dimensionality of the irreducible representation with graph $[\tau]$. The quantities $\hat{\rho}^{[\tau]}_{rs}$ are usually referred to as <u>Young operators</u>. Explicit matrices for the so-called Young-Yamanouchi standard orthogonal representation, for k = 2,3,4,5,6 can be found in ref.[13].

In the case of the one-row diagram $[k]$ of S_k, there is only one standard tableaux $\boxed{1}\boxed{2}\boxed{3}\cdots\boxed{k}$ and the representation is one-dimensional, with $\Gamma^{[k]}(\hat{P}_j) = 1, \forall \hat{P}_j$. The corresponding Young operator $\hat{\rho}^{[k]} = (1/k!)\sum_{j=1}^{k!} \hat{P}_j$ is the projector for the totally symmetric representation Γ^s previously introduced; hence, $\Gamma^s \equiv \Gamma^{[k]}$. Passing to the one-column diagram $[1^k]$ of S_k, the representation is one-dimensional as well, with $\Gamma^{[1^k]} = \pm 1$, the upper (lower) sign being appropriate for even (odd) permutations; the corresponding Young operator

$$\hat{\rho}^{[1^k]} \equiv \hat{\mathcal{A}} = (1/k!) \sum_{j=1}^{k!} (-1)^{p_j} \hat{P}_j \tag{II.3.2}$$

is the antisymmetrizer, i.e. the projector associated with the fully antisymmetric representation previously called Γ^a, and $\Gamma^a = \Gamma^{[1^k]}$.

Starting from the Young operators, we can operate on general functions, so as to generate sets of basis functions which transform according to given irreducible representations of the symmetric group S_k.

II.4. Pauli Principle Implications.

The Hamiltonian operators we shall be concerned with do not explicitly involve electron spin operators (spin free Hamiltonian). In spite of this, the electron spin plays a fundamental role through Pauli principle, by restricting the possible kinds of symmetries allowable to purely spatial functions.

To ascertain this very important point, we recall that Pauli principle selects as only allowed state functions $\Psi(\underline{x}_1,\ldots,\underline{x}_N)$ for a N electron system those which are fully antisymmetric under the interchange of the space-spin coordinates $\underline{x}_i \equiv (\underline{r}_i,\sigma_i)$ of any two electrons. The only permissible quantum states are therefore those described by wavefunctions which transform according to the fully antisymmetric representation under the group of operators $\hat{P}_i$ which permute the space-spin coordinates of the N electrons.

Let $\varphi_j(\underline{r}_1,\ldots,\underline{r}_N)$, $j = 1,\ldots,\lambda$, be a set of degenerate spatial eigenstates to the spin-free Hamiltonian $\hat{H}$, corresponding to the energy value E. Since $\hat{H}$ is spin-independent, any φ_j can be multiplied by an arbitrary function $\chi_i\ (\sigma_1,\ldots,\sigma_N)$ of the spin coordinates to give a product function $\varphi_j\ (\underline{r}_1,\ldots,\underline{r}_N)\ \chi_i\ (\sigma_1,\ldots,\sigma_N)$ which is still an eigenstate to $\hat{H}$ with energy E; moreover, any linear combination

$$\Psi = \sum_{ij} \varphi_j\ (\underline{r}_1,\ldots,\underline{r}_N)\ \chi_i\ (\sigma_1,\ldots,\sigma_N)$$

will be eigenstate to $\hat{H}$, with the same energy E.

From the theorem explicitly enunciated in the preceding section we have learnt that only the direct product of two associate representations will contain the totally antisymmetric representation of S_N. Assuming that the degenerate spatial eigenstates $\varphi_j^{[\tau]}$, $j = 1,\ldots,\lambda_\tau$, belong to some irreducible representation $\Gamma^{[\tau]}$ of S_N, only if we are able to build up spin functions $\chi_i^{[\tilde{\tau}]}$, $i = 1,\ldots,\lambda_\tau$, which transform under spin coordinate permutations according to the representation $\Gamma^{[\tilde{\tau}]}$ associated to $\Gamma^{[\tau]}$, we may obtain by linear combination of the products $\varphi_j^{[\tau]}\chi_i^{[\tilde{\tau}]}$ an eigenstate $\Psi^{[1^N]}$ which is Pauli allowed, because fully antisymmetric in the space-spin coordinates of all electrons. Since the λ_τ^2-dimensional space spanned by $\varphi_j^{[\tau]}\chi_i^{[\tilde{\tau}]}$ contains one and only one antisymmetric function, in order to find it we may start with any product of such space and then to project out the totally antisymmetric function by using (II.3.2):

$$\Psi^{[1^N]}(\underline{x}_1,\ldots,\underline{x}_N) = (1/N!)\sum_{j=1}^{N!}(-1)^{p_j}\ \hat{P}_j\ \varphi_1^{[\tau]}\ (\underline{r}_1,\ldots,\underline{r}_N)\ \chi_1^{[\tilde{\tau}]}(\sigma_1,\ldots,\sigma_N)$$

($\hat{P}_j$ operates on both space and spin coordinates).

Therefore we can conclude that Pauli principle allows spatial functions corresponding to the Young diagram $[\tau]$ depending on whether it is permissible to build spin functions belonging to the associate diagram $[\tilde{\tau}]$.

Now the spin space of a N electron system is spanned by 2^N basis functions, all the possible products of one-electron spin eigenstates $\alpha(\sigma_i)$ or $\beta(\sigma_i)$, as follows

$\alpha(\sigma_1)\alpha(\sigma_2)\ldots\alpha(\sigma_N)$, $\beta(\sigma_1)\alpha(\sigma_2)\ldots\alpha(\sigma_N)$, $\alpha(\sigma_1)\beta(\sigma_2)\ldots\alpha(\sigma_N)$, ..., $\beta(\sigma_1)\beta(\sigma_2)\ldots\beta(\sigma_N)$.

Such a space is clearly invariant under the set of N! permutations which interchange the electron spin coordinates σ_i, so that we should be able to split it into subspaces belonging to the various irreducible representations of the symmetric group S_N. However, it can be verified that it is not possible to build up spin functions whose symmetry corresponds to Young diagrams having more than two rows; consequently, Pauli principle allows only spatial symmetries corresponding to graphs having two columns at most.

For two electrons there are the following possibilities

space spin and space spin

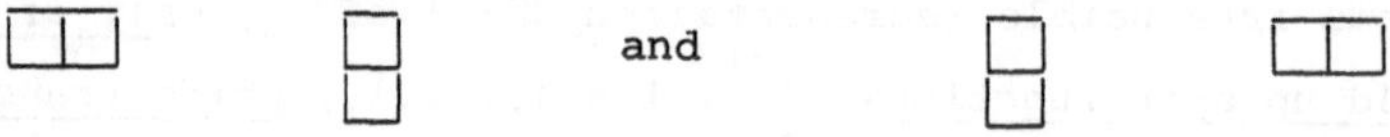

i.e. no restriction on the possible kinds of space symmetry. On the other hand, passing to four electrons, we have

space spin , space spin , space spin

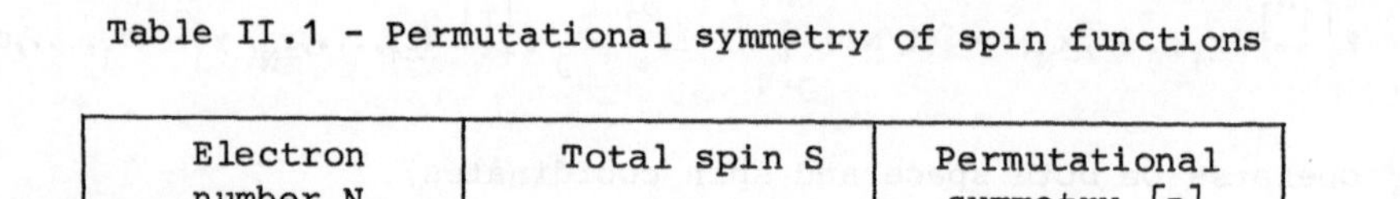

i.e. the space graphs 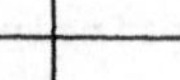 and  are missing.

Table II.1 - Permutational symmetry of spin functions

Electron number N	Total spin S	Permutational symmetry $[\tau]$
2	0 1	$[1^2]$ $[2]$
3	1/2 3/2	$[2,1]$ $[3]$
4	0 1 2	$[2^2]$ $[3,1]$ $[4]$

In Table II.1 are reported for some simple electron systems the correspondences between (total) spin eigenvalues (as result, for instance, from the branching diagram [7]) and the permutational symmetry of the associated spin functions [14]. We note that, in general, the Young diagram $[\tau] \equiv [\tau_1,\tau_2]$ corresponds to the spin value $S = (1/2)(\tau_1 - \tau_2)$; conversely, from $\tau_1 + \tau_2 = N$, the number of electrons, a system with total spin eigenvalue S has a permutational symmetry $[\tau] = [\tau_1,\tau_2]$, where $\tau_1 = \frac{N}{2} + S$, $\tau_2 = \frac{N}{2} - S$; graphically

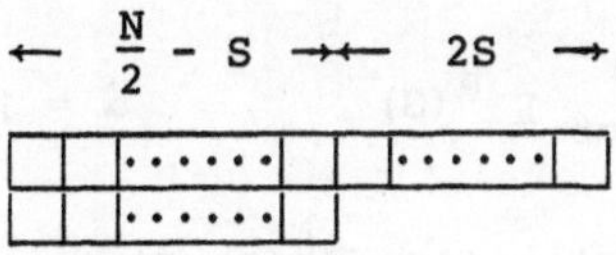

II.5. Induced and Subduced Representations.

The concepts of *induced* and *subduced representations* [15] will be found useful working tools in Ch.IV, so we provide here a short outline of the subject: in order to keep the level as concrete as possible, our presentation will be founded on the familiar use of basis functions, instead of the more abstract language of the group algebra.

Let us consider a set of functions $\{\varphi_1,..,\varphi_{\lambda'}\}$ which are a basis for a λ'-dimensional irreducible representation $\gamma(G')$ of the subgroup G' of a group G and suppose that the action on these functions of any element of G, not contained in G', gives a result outside the space they span. If we then apply all the elements of G to the set $\{\varphi_1,..\varphi_{\lambda'}\}$, we shall generate a larger set of independent functions $\{\varphi_1,..\varphi_\lambda\}$, $\lambda > \lambda'$, which affords a representation of G, called the *representation induced in G by the (irreducible) representation $\gamma(G')$ of the subgroup G'* and denoted as $\gamma(G')\uparrow G$.

Through an inverse procedure it is also possible, starting from a given representation of the group G, to generate a representation of any subgroup G' of G itself. If $\Gamma(G)$ is a representation of the group G, we may *subduce* a representation $\Gamma(G')$ of the subgroup G' by simply taking the subset of representative matrices $\Gamma(\hat{g}_1')$, $\Gamma(\hat{g}_2')$,..... from the set $\Gamma(G)$, corresponding to the elements $\hat{g}_1', \hat{g}_2', \ldots$ of the subgroup G'. Such a representation, generally reducible, is named <u>representation subduced in G' by the representation $\Gamma(G)$ of the full group G</u> and denoted as $\Gamma(G)\downarrow G'$.

An important theorem due to Frobenius, known as <u>Reciprocity Theorem</u>, relates the two procedures described above. If we use the notation $\gamma^{\alpha}(G')$ for the irreducible representations of G', while $\Gamma^{\beta}(G)$ indicates the irreducible representations of the full group G, the reduction series for $\gamma^{\alpha}(G')\uparrow G$ will be given by

$$\gamma^{\alpha}(G')\uparrow G = \sum_{\beta=1}^{n_G} a_{\alpha\beta}\,\Gamma^{\beta}(G) \quad , \quad \alpha = 1,\ldots,n_G \ .$$

The analogous reduction series for $\Gamma^{\beta}(G)\downarrow G'$ is

$$\Gamma^{\beta}(G)\downarrow G' = \sum_{\alpha=1}^{n'_{G'}} b_{\beta\alpha}\,\gamma^{\alpha}(G') \quad , \quad \beta = 1,\ldots,n'_{G'} \ ,$$

where n_G is the number of irreducible representations of the group G and $n'_{G'}$ the corresponding number for the irreducible representations of the subgroup G'. The Reciprocity Theorem then states that

$$a_{\alpha\beta} = b_{\beta\alpha} \ , \quad \alpha = 1,\ldots,n_G \ ; \ \beta = 1,\ldots,n'_{G'} \ .$$

As an example of application of this theorem [17], let us consider the symmetric group S_4 and its subgroup $S_2 \otimes \bar{S}_2 \equiv \{\hat{1},(12)\} \otimes \{\hat{1},(34)\} = \{\hat{1},(12),(34),(12)(34)\}$. The irreducible representations of the group S_2 correspond to the one-dimensional Young graphs [2] and $[1^2]$, where the transposition is represented by +1 and −1, respectively. We have therefore the following character table for the subgroup $S_2 \otimes \bar{S}_2$:

$S_2 \otimes \bar{S}_2$	$\hat{1}$	(12)	(34)	(12)(34)
$[2] \otimes [2]$	1	1	1	1
$[2] \otimes [1^2]$	1	1	−1	−1
$[1^2] \otimes [2]$	1	−1	1	−1
$[1^2] \otimes [1^2]$	1	−1	−1	1

The characters of the same elements in the various irreducible representations of S_4 are (see, for instance, ref. [13]):

S_4	$\hat{1}$	(12)	(34)	(12)(34)
[4]	1	1	1	1
[3,1]	3	1	1	-1
[2^2]	2	0	0	2
[2,1^2]	3	-1	-1	-1
[1^4]	1	-1	-1	1

From the basic character theorem the following reduction series for the subduced representations $\Gamma^{\beta}(S_4)\downarrow S_2 \otimes \bar{S}_2$ result:

$$[4]\downarrow S_2\otimes\bar{S}_2 = [2]\otimes[2]$$
$$[3,1]\downarrow S_2\otimes\bar{S}_2 = [2]\otimes[2] + [2]\otimes[1^2] + [1^2]\otimes[2]$$
$$[2^2]\downarrow S_2\otimes\bar{S}_2 = [2]\otimes[2] + [1^2]\otimes[1^2]$$
$$[2,1^2]\downarrow S_2\otimes\bar{S}_2 = [2]\otimes[1^2] + [1^2]\otimes[2] + [1^2]\otimes[1^2]$$
$$[1^4]\downarrow S_2\otimes\bar{S}_2 = [1^2]\otimes[1^2]$$

These decompositions tell us that it is possible to choose for the S_4 representations basis functions affording at the same time the representations of $S_2\otimes\bar{S}_2$ which occur in the subduced representations, so that we have functions simultaneously adapted to the symmetry of $S_2\otimes\bar{S}_2$ and S_4.

From the Reciprocity Theorem we then get the following representation for the induced representations $\gamma^{\alpha}(S_2\otimes S_2)\uparrow S_4$:

$$[2]\otimes[2]\uparrow S_4 = [4] + [3,1] + [2^2]$$
$$[2]\otimes[1^2]\uparrow S_4 = [3,1] + [2,1^2]$$
$$[1^2]\otimes[2]\uparrow S_4 = [3,1] + [2,1^2]$$
$$[1^2]\otimes[1^2]\uparrow S_4 = [2^2] + [2,1^2] + [1^4]$$

All the previous results can be collected as in Table II.2, where the entry belonging to a given row and column simultaneously provides the number of occurrences of $\gamma^{\alpha}(S_2 \otimes \bar{S}_2)$ in $\Gamma^{\beta}(S_4) \downarrow S_2 \otimes \bar{S}_2$ and $\Gamma^{\beta}(S_4)$ in $\gamma^{\alpha}(S_2 \otimes \bar{S}_2) \uparrow S_4$.

Table II.2 - Number of occurrences of $\gamma^{\alpha}(S_2 \otimes \bar{S}_2)$ in $\Gamma^{\beta}(S_4) \downarrow S_2 \otimes \bar{S}_2$ and $\Gamma^{\beta}(S_4)$ in $\gamma^{\alpha}(S_2 \otimes \bar{S}_2) \uparrow S_4$ (Reciprocity Theorem).

$S_2 \otimes \bar{S}_2$ \ S_4	[4]	[3,1]	[2^2]	[2,1^2]	[1^4]
[2]⊗[2]	1	1	1	0	0
[2]⊗[1^2]	0	1	0	1	0
[1^2]⊗[2]	0	1	0	1	0
[1^2]⊗[1^2]	0	0	1	1	1

III. SYMMETRY-ADAPTED PERTURBATION THEORY: A GENERAL APPROACH

III.1. Eigenvalue Problems and Partitioning Technique.

Perturbation theory plays a fundamental role in the determination of quantum mechanical states. Here we shall be concerned with stationary states, so our attention will be restricted only to the time-independent variant of such theory. Besides conventional treatments, described in any textbook on quantum mechanics, a renewed interest in general perturbation theory is to be recognized in more recent years [18,19]. In a remarkable series of papers, entitled "Studies in Perturbation Theory" [10,20-32], Löwdin has emphasized the partitioning technique as a tool for solving secular equations associated with eigenvalue problems, putting in evidence its connection with various perturbation expansions.

We now propose to review the partitioning technique in some detail, because its generality lends itself to generate symmetry-adapted perturbative schemes (to be better appreciated in due time), by which the non-trivial problem of incorporating into perturbation theory the treatment of the constants of motion and the symmetry properties of the system can be approached. We shall make use of an operator formalism, since it appears particularly powerful, compact and elegant.

In order to solve the Schrödinger equation

$$\hat{H}|\Psi\rangle = E|\Psi\rangle \qquad \text{(III.1.1)}$$

we start by partitioning the Hilbert space in two subspaces: a reference space, spanned by the vector $|\varphi\rangle$, and the complementary space. A projection operator $\hat{O}$ is associated with $|\varphi\rangle$, i.e.

$$\hat{O} = |\varphi\rangle\langle\varphi|\varphi\rangle^{-1}\langle\varphi| \qquad \text{(III.1.2)}$$

(obviously, $\hat{O}^\dagger = \hat{O}$, $\hat{O}^2 = \hat{O}$); a projection operator $\hat{P}$ can analogously be associated with the orthogonal complement to $|\varphi\rangle$,

$$\hat{P} = \hat{1} - \hat{O} \qquad \text{(III.1.3)}$$

and $\hat{P}^\dagger = \hat{P}$, $\hat{P}^2 = \hat{P}$, $\hat{P}\hat{O} = \hat{O}\hat{P} = 0$. Clearly $\hat{O}|\varphi\rangle = |\varphi\rangle$, $\hat{P}|\varphi\rangle = 0$.

Let us now consider the <u>auxiliary Hamiltonian</u> operator $\hat{H}$,

$$\hat{H} = \hat{P}\hat{H}\hat{P} \qquad \text{(III.1.4)}$$

It has $|\varphi\rangle$ as a trivial eigenstate, corresponding to the eigenvalue zero; all other eigenstates lie in the orthogonal complement to $|\varphi\rangle$ and have eigenvalues bracketed by those of $\hat{H}$ in order [10].

The resolvent $\hat{R}$ of the auxiliary Hamiltonian, $\hat{R}(\varepsilon) = (\varepsilon - \hat{H})^{-1}$ (ε is an arbitrary complex number), exhibits poles at the real eigenvalues of $\hat{H}$, but not for $\varepsilon = E_k$, the eigenvalues of the Hamiltonian operator $\hat{H}$.

The fundamental quantity in the partitioning technique is

$$\hat{T}(\varepsilon) = \hat{R}(\varepsilon)\hat{P} = (\varepsilon - \hat{H})^{-1}\hat{P}, \qquad \text{(III.1.5)}$$

which is called <u>reduced resolvent</u>. If we note that $\hat{H}\hat{P} = \hat{P}\hat{H}$, so that $\hat{P}(\varepsilon - \hat{H})^{-1} = (\varepsilon - \hat{H})^{-1}\hat{P}$, it follows that

$$\hat{T} = \hat{T}\hat{P} = \hat{P}\hat{T} \qquad \text{(III.1.6)}$$

and then the basic result

$$\hat{P}(\varepsilon - \hat{H})\hat{T} = \hat{P} \qquad \text{(III.1.7)}$$

Let us now introduce the following trial eigenvector $|\Psi(\varepsilon)\rangle$,

$$|\Psi(\varepsilon)\rangle = |\varphi\rangle + |\phi(\varepsilon)\rangle = |\varphi\rangle + \hat{T}(\varepsilon)\hat{H}|\varphi\rangle \qquad \text{(III.1.8)}$$

Since $|\phi(\varepsilon)\rangle$ lies in the orthogonal complement to $|\varphi\rangle$, i.e. $\langle\varphi|\phi(\varepsilon)\rangle \equiv 0$, $|\Psi(\varepsilon)\rangle$ verifies the so-called "intermediate normalization" condition

$$\langle\varphi|\Psi(\varepsilon)\rangle = \langle\varphi|\varphi\rangle \qquad \text{(III.1.9)}$$

If we let $\hat{P}(\hat{H} - \varepsilon)$ to operate on both sides of eq. (III.1.8), it follows $\hat{P}(\hat{H} - \varepsilon)|\Psi(\varepsilon)\rangle = 0$, i.e.

$$(\hat{H} - \varepsilon)|\Psi(\varepsilon)\rangle = \hat{O}(\hat{H} - \varepsilon)|\Psi(\varepsilon)\rangle = |\varphi\rangle \frac{\langle\varphi|\hat{H} - \varepsilon|\Psi(\varepsilon)\rangle}{\langle\varphi|\varphi\rangle} =$$

$$= |\varphi\rangle\left[\frac{\langle\varphi|\hat{H} + \hat{H}\hat{T}(\varepsilon)\hat{H}|\varphi\rangle}{\langle\varphi|\varphi\rangle} - \varepsilon\right] \qquad \text{(III.1.10)}$$

Therefore the "trial eigenvector" $|\Psi(\varepsilon)\rangle$ satisfies an <u>inhomogeneous</u> Schrödinger equation, which drops back into the eigenvalue problem (III.1.1) only when

$$\varepsilon = \frac{\langle\varphi|\hat{H} + \hat{H}\hat{T}(\varepsilon)\hat{H}|\varphi\rangle}{\langle\varphi|\varphi\rangle} = E \qquad \text{(III.1.11)}$$

The eigenvalues contained in the reference space can, in principle, be obtained by solving the implicit equation (III.1.11), $\varepsilon = f(\varepsilon)$, where we have put $f(\varepsilon) = \langle\varphi|\varphi\rangle^{-1}\langle\varphi|\hat{H} + \hat{H}\hat{T}(\varepsilon)\hat{H}|\varphi\rangle$.

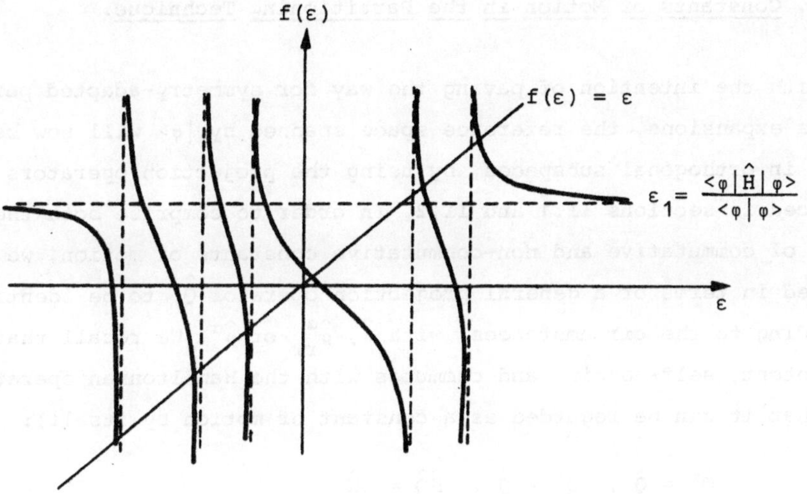

Figure III.1. - The "bracketing function" $f(\varepsilon)$.

A schematic behaviour of the function $f(\varepsilon)$ is shown in Fig. III.1. The branches of $f(\varepsilon)$ are best understood by making recourse to a spectral representation of the reduced resolvent in terms of the eigenstates to $\hat{\mathcal{H}} = \hat{\mathcal{P}}\hat{H}\hat{\mathcal{P}}$. The vertical asymptotes are in correspondence with the eigenvalues of the auxiliary Hamiltonian $\hat{\mathcal{H}}$ (which are not simultaneous eigenvalues to $\hat{H}$)*. The exact eigenvalues E to $\hat{H}$ correspond to the

*In case there exist eigenvectors $|E_i\rangle$ to $\hat{H}$ which are entirely contained in the orthogonal complement to $|\varphi\rangle$, i.e. $\langle\varphi|E_i\rangle = 0$, they are simultaneous eigenvectors to $\hat{\mathcal{H}}$, with identical eigenvalues E_i. The spectral decomposition of $f(\varepsilon)$, $f(\varepsilon) = \langle\varphi|\varphi\rangle^{-1}\{\langle\varphi|\hat{H}|\varphi\rangle + \Sigma_k|\langle\varphi|H|\bar{E}_k\rangle|^2/(\varepsilon - \bar{E}_k)\}$, where $\hat{\mathcal{H}}|\bar{E}_k\rangle = \bar{E}_k|\bar{E}_k\rangle$, is then characterized by the lack of the corresponding vertical asymptotes, because $\langle\varphi|\hat{H}|E_i\rangle = 0$.

crossing-points between the curve $y = f(\varepsilon)$ and the straight line $y = \varepsilon$ and could be approached by searching the zero-points of the quantity $\varepsilon - f(\varepsilon)$. The figure makes evident by itself the so-called "bracketing theorem" [22,28,29], which states that the two numbers $(\varepsilon, f(\varepsilon))$ bracket at least one exact eigenvalue E. Because of this property, the function $f(\varepsilon)$ is referred to as "bracketing function".

III.2. Constants of Motion in the Partitioning Technique.

With the intention of paving the way for symmetry-adapted perturbative expansions, the reference space spanned by $|\varphi\rangle$ will now be split in orthogonal subspaces, by using the projection operators introduced in sections II.1 and II.2. In order to comprise both the cases of commutative and non-commutative constants of motion, we shall proceed in terms of a general projection operator $\hat{Q}$, to be identified according to the circumstances, with $\hat{\Lambda}$, $\hat{\rho}^{\alpha}_{rr}$ or $\hat{\rho}^{\alpha}$. We recall that $\hat{Q}$ is idempotent, self-adjoint and commutes with the Hamiltonian operator $\hat{H}$ (so that it can be regarded as a constant of motion by itself):

$$\hat{Q}^2 = \hat{Q} \ , \quad \hat{Q}^{\dagger} = \hat{Q} \ , \quad \hat{H}\hat{Q} = \hat{Q}\hat{H}$$

Let us now introduce the new (explicitly normalized) reference state vector $|\varphi'\rangle = \hat{Q}|\varphi\rangle\langle\varphi|\hat{Q}|\varphi\rangle^{-1/2}$ and the associated projection operator

$$\hat{O} = |\varphi'\rangle\langle\varphi'| = \hat{Q}|\varphi\rangle\langle\varphi|\hat{Q}|\varphi\rangle^{-1}\langle\varphi|\hat{Q} \qquad \text{(III.2.1)}$$

(obviously, $\hat{O}^2 = \hat{O}$, $\hat{O}^{\dagger} = \hat{O}$). The projector for the orthogonal complement to $|\varphi'\rangle$, $\hat{P} = \hat{1} - \hat{O}$, satisfies $\hat{P}^2 = \hat{P}$, $\hat{P}^{\dagger} = \hat{P}$, $\hat{P}\hat{O} = \hat{O}\hat{P} = 0$ and $\hat{P}|\varphi'\rangle = 0$.

In order to deal with the specific subspace generated by $\hat{Q}$, it is convenient to define the projected Hamiltonian $\hat{H}_Q$,

$$\hat{H}_Q = \hat{Q}\hat{H}\hat{Q} = \hat{H}\hat{Q} = \hat{Q}\hat{H} \qquad \text{(III.2.2)}$$

and the auxiliary Hamiltonian operator $\hat{H}_1 = \hat{P}\hat{H}_Q\hat{P} = (\hat{P}\hat{Q})\hat{H}(\hat{Q}\hat{P})$. The operator $\hat{P}_1 = \hat{P}\hat{Q}$ is easily seen to satisfy the relations

$$\hat{P}_1 = \hat{P}\hat{Q} = \hat{Q}\hat{P} = \hat{Q}\hat{P}\hat{Q} = \hat{Q} - \hat{O} \qquad \text{(III.2.3)}$$

If we furthermore introduce the new (not self-adjoint) operator

$$\hat{\omega} = \hat{1} - |\varphi\rangle\langle\varphi|\hat{Q}|\varphi\rangle^{-1}\langle\varphi|\hat{Q} \qquad \text{(III.2.4)}$$

we can also write $\hat{P}_1$ as

$$\hat{P}_1 = \hat{Q}\hat{\omega} = \hat{\omega}^{\dagger}\hat{Q} \qquad \text{(III.2.5)}$$

Obviously $\hat{\omega}|\varphi\rangle = \hat{P}_1|\varphi\rangle = 0$, $\hat{P}_1^{\dagger} = \hat{P}_1$, $\hat{P}_1^2 = \hat{P}_1$, $\hat{\omega}^2 = \hat{\omega}$, $(\hat{\omega}^{\dagger})^2 = (\hat{\omega}^{\dagger})$. We learn therefore that $\hat{P}_1$ is a self-adjoint projection operator, associated with the orthogonal complement to $\hat{O}$ <u>with respect to the subspace generated by</u> $\hat{Q}$.

We now intend to reduce all the quantities of importance involved in the partitioning technique to such subspace, instead of the whole Hilbert space. From the reduced resolvent $\hat{T}(\varepsilon)$, eq. (III.1.5), we define a new reduced resolvent operator $\hat{T}_1(\varepsilon)$, which represents the component of $\hat{T}(\varepsilon)$ in the given subspace:

$$\hat{T}_1(\varepsilon) = \hat{Q}\hat{T}(\varepsilon)\hat{Q} = \hat{T}(\varepsilon)\hat{Q} = \hat{Q}\hat{T}(\varepsilon) \qquad \text{(III.2.6)}$$

Now $\hat{H}_1 = \hat{P}_1 H \hat{P}_1 = \hat{Q}\hat{H}\hat{Q} = \hat{Q}\hat{H} = \hat{H}\hat{Q}$, where $\hat{H} = \hat{P}\hat{H}\hat{P}$; taking into account that $\hat{P}_1\hat{Q} = \hat{Q}\hat{P}_1 = \hat{P}_1$, we obtain

$$\hat{P}_1(\varepsilon - \hat{H}) = \hat{P}_1\hat{Q}(\varepsilon - \hat{H}) = \hat{P}_1(\varepsilon - \hat{H}_1) = (\varepsilon - \hat{H}_1)\hat{P}_1$$

and therefore

$$(\varepsilon - \hat{H}_1)^{-1}\hat{P}_1 = \hat{P}_1(\varepsilon - \hat{H})^{-1} \qquad \text{(III.2.7)}$$

After a few manipulations it follows that

$$\hat{T}_1(\varepsilon) = (\varepsilon - \hat{H}_1)^{-1}\hat{P}_1 \qquad \text{(III.2.8)}$$

Proceeding in complete analogy with sec. III.1, we shall now introduce the "trial eigenvector" $|\Psi(\varepsilon)\rangle$

$$|\Psi(\varepsilon)\rangle = |\varphi'\rangle + \hat{T}_1(\varepsilon)\hat{H}|\varphi'\rangle = \hat{Q}\frac{[\hat{1} + \hat{T}_1(\varepsilon)\hat{H}]}{\langle\varphi|\hat{Q}|\varphi\rangle^{1/2}}|\varphi\rangle \qquad \text{(III.2.9)}$$

As $\hat{Q}|\psi(\varepsilon)\rangle = |\psi(\varepsilon)\rangle$, the "trial eigenvector" entirely lies in the subspace generated by $\hat{Q}$; it satsfies the inhomogeneous Schrödinger equation

$$(\hat{H} - \varepsilon)|\psi(\varepsilon)\rangle = |\varphi'\rangle[\langle\varphi'|\hat{H} + \hat{H}\hat{T}_1(\varepsilon)\hat{H}|\varphi'\rangle - \varepsilon] \qquad \text{(III.2.10)}$$

and fulfills the intermediate normalization condition

$$\langle\varphi'|\Psi(\varepsilon)\rangle = \langle\varphi'|\varphi'\rangle = 1$$

The inhomogeneous Schrödinger equation can be rewritten in terms of the primitive reference function $|\varphi\rangle$:

$$(\hat{H} - \varepsilon)|\Psi(\varepsilon)\rangle = \frac{\hat{Q}|\varphi\rangle}{\langle\varphi|\hat{Q}|\varphi\rangle^{\frac{1}{2}}}\left[\frac{\langle\varphi|\hat{H}\hat{Q} + \hat{H}\hat{T}_1(\varepsilon)\hat{H}|\varphi\rangle}{\langle\varphi|\hat{Q}|\varphi\rangle} - \varepsilon\right] \qquad \text{(III.2.11)}$$

that defines the new "bracketing function"

$$\varepsilon_1 = f(\varepsilon) = \frac{\langle\varphi|\hat{H}\hat{Q} + \hat{H}\hat{T}_1(\varepsilon)\hat{H}|\varphi\rangle}{\langle\varphi|\hat{Q}|\varphi\rangle} \qquad \text{(III.2.12)}$$

for which the general properties sketched in sec. III.1 are clearly to be expected.

We can therefore conclude that it is possible to consider one subspace belonging to a given projector operator $\hat{Q}$ at a time and to apply the partitioning technique to it, that usually amounts to a noticeable simplification of the complexity of the problem at issue.

Symmetry-adapted perturbative expansions of the energy can be obtained starting from eq. (III.2.12): this will be the subject of the next section.

III.3. Connection between Partitioning Technique and Perturbation Theory.

We now introduce the standard assumption of perturbation theory, that the Hamiltonian operator $\hat{H}$ can be split in the form

$$\hat{H} = \hat{H}_0 + \hat{V} \qquad \text{(III.3.1)}$$

where $\hat{H}_0$ is the unperturbed Hamiltonian and $\hat{V}$ the perturbation operator.

The partition expressed by eq. (III.3.1) is often dictated by physical arguments, such as in the case of atoms or molecules subjected to external fields. In this kind of situations, one observes that the transition from $\hat{H}_0$ to $\hat{H}$ is associated with a lowering of simmetry of the system, so that any constant of motion to $\hat{H}$ is also a constant of motion to $\hat{H}_0$; hence, $[\hat{H},\hat{Q}] = [\hat{H}_0,\hat{Q}] = 0$, for all the projectors $\hat{Q}$ associated with the constants of motion to $\hat{H}$.

The problems we shall mainly be interested in the present notes require that we relax such a restraint, the transition from $\hat{H}_0$ to $\hat{H}$ involving a possible complete change of symmetry of the system. This is, for instance, the case of the permutational symmetry, i.e. the symmetry associated with the electron exchange, as we shall fully discuss in the next chapter.

From $\hat{H}\hat{Q} = \hat{Q}\hat{H}$ and eq. (III.3.1) we get

$$[\hat{H}_0,\hat{Q}] = [\hat{Q},\hat{V}] \qquad \text{(III.3.2)}$$

a general consequence of the fact that a constant of motion to $\hat{H}$ is not necessarily a constant of motion to $\hat{H}_0$.

Let then $|\varphi_0\rangle$ denote the eigenstate corresponding to the non-degenerate unperturbed level E_0, i.e.

$$\hat{H}_0|\varphi_0\rangle = E_0|\varphi_0\rangle \qquad \text{(III.3.3)}$$

If we choose $|\varphi_0\rangle$ as the reference vector $|\varphi\rangle$ in eq. (III.2.12), we obtain

$$\varepsilon_1 = f(\varepsilon) = E_0 + \frac{\langle\varphi_0|\hat{V}\hat{Q} + \hat{V}\hat{T}_1(\varepsilon)\hat{V}|\varphi_0\rangle}{\langle\varphi_0|Q|\varphi_0\rangle} \qquad \text{(III.3.4)}$$

because $\hat{P}_1|\varphi_0\rangle = \hat{T}_1(\varepsilon)|\varphi_0\rangle = 0$.

In order to proceed further, a suitable form for the reduced resolvent $\hat{T}_1(\varepsilon)$ is needed, which is susceptible of perturbative expansion in orders of perturbation $\hat{V}$. To begin with, we note that $\hat{H}_1 = \hat{P}_1\hat{H}\hat{P}_1 = \hat{\omega}^{\dagger}\hat{Q}\hat{H}\hat{Q}\hat{\omega} = \hat{\omega}^{\dagger}\hat{H}\hat{Q}\hat{\omega} = \hat{\omega}^{\dagger}\hat{H}\hat{P}_1$, so that $\hat{H}_1 = \hat{\omega}^{\dagger}\hat{H}_0\hat{P}_1 + \hat{\omega}^{\dagger}\hat{V}\hat{P}_1$. But from eqs. (III.2.4) and (III.3.3) it follows

$$\hat{\omega}^{\dagger}\hat{H}_0\hat{P}_1 = \hat{H}_0\hat{P}_1 \qquad \text{(III.3.5)}$$

and therefore

$$\hat{H}_1 = (\hat{H}_0 + \hat{\omega}^\dagger\hat{V})\hat{P}_1 \tag{III.3.6}$$

Let us now reconsider the expression for $\hat{T}_1(\varepsilon)$, eq. (III.2.8), in the form $\hat{T}_1(\varepsilon) = [a - H_1 - (a - \varepsilon)]^{-1}P_1$, where a is an arbitrary complex number. Taking into account eq. (III.3.6), we also have

$$\begin{aligned}\hat{T}_1(\varepsilon) &= [a - \hat{H}_0\hat{P}_1 - \hat{\omega}^\dagger\hat{V}\hat{P}_1 - (a - \varepsilon)\hat{\omega}^\dagger\hat{P}_1]^{-1}\hat{P}_1 = \\ &= [(a - \hat{H}_0) - \hat{\omega}^\dagger\hat{V}']^{-1}\hat{P}_1 = \\ &= [\hat{1} - (a - \hat{H}_0)^{-1}\hat{\omega}^\dagger\hat{V}']^{-1}(a - \hat{H}_0)^{-1}\hat{\omega}^\dagger\hat{P}_1\end{aligned} \tag{III.3.7}$$

where

$$\hat{V}' = \hat{V} - (\varepsilon - a) \tag{III.3.8}$$

If we then introduce the following reduced resolvent associated with the unperturbed Hamiltonian,

$$\hat{T}_0(a) = [a - \hat{H}_0]^{-1}[\hat{1} - |\varphi_0\rangle\langle\varphi_0|] \tag{III.3.9}$$

it is easily verified that $\hat{T}_0(a)\hat{\omega}^\dagger = [a - \hat{H}_0]^{-1}\hat{\omega}^\dagger$, because $\hat{\omega}|\varphi_0\rangle = 0$. Eq. (III.3.7) can hence be cast into the basic form

$$\hat{T}_1(\varepsilon) = [\hat{1} - \hat{T}_0(a)\hat{\omega}^\dagger\hat{V}']^{-1}\hat{T}_0(a)\hat{P}_1 \tag{III.3.10}$$

which involves the unperturbed reduced resolvent $\hat{T}_0(a)$.

The expression (III.3.10) is a compact, closed form for $\hat{T}_1(\varepsilon)$, which is necessarily independent of the value chosen for the parameter a; moreover, from $\hat{T}_1(\varepsilon) = \hat{P}_1(\varepsilon - H_1)^{-1}\hat{P}_1$, one concludes that $\hat{T}_1(\varepsilon) = \hat{T}_1^\dagger(\varepsilon)$, $\hat{T}_1(\varepsilon) = \hat{P}_1\hat{T}_1(\varepsilon) = \hat{T}_1(\varepsilon)\hat{P}_1$, i.e. $\hat{T}_1(\varepsilon)$ is self-adjoint and completely contained in the subspace generated by the projector $\hat{P}_1$.

Actual applications of eq. (III.3.10) almost invariably involve various steps of approximation; if we confine ourselves to perturbative procedures, the expansion of the inverse operator occurring in eq. (III.3.10) will be the standard approach, followed, as a practical rule, by the successive truncation of the resulting series after the first few terms.

The expansion of the operator quantity $(\hat{A} - \hat{B})^{-1}$ can be carried out by iterating one of the following operator identities

$$(\hat{A} - \hat{B})^{-1} = \hat{A}^{-1} + \hat{A}^{-1}\hat{B}(\hat{A} - \hat{B})^{-1}$$
$$(\hat{A} - \hat{B})^{-1} = \hat{A}^{-1} + (\hat{A} - \hat{B})^{-1}\hat{B}\hat{A}^{-1} \qquad \text{(III.3.11)}$$

or their symmetric combination,

$$(\hat{A} - \hat{B})^{-1} = \hat{A}^{-1} + \hat{A}^{-1}\hat{B}\hat{A}^{-1} + \hat{A}^{-1}\hat{B}(\hat{A} - \hat{B})^{-1}\hat{B}\hat{A}^{-1} \qquad \text{(III.3.12)}$$

We remark that, although such a procedure is quite legitimate, the individual terms arising from the manipulation of eq. (III.3.10) are not self-adjoint and are not situated in the subspace generated by $\hat{P}_1$. If one searches an expansion characterized by such features, a more involved procedure is needed. The first step is constituted by the use of an expansion for the reduced resolvent that is authomatically self-adjoint, followed by a second step involving a projection onto the desired subspace. If $\hat{T}$ is an arbitrary operator having the general structure

$$\hat{T} = (\hat{1} - \hat{B})^{-1}\hat{C} \qquad \text{(III.3.13)}$$

it can be shown that the identity [10]

$$\hat{T} = \frac{1}{2}(\hat{C} + \hat{C}^{\dagger}) + \frac{1}{2}(\hat{B}\hat{C}^{\dagger} + \hat{C}\hat{B}^{\dagger}) + \hat{B}\hat{T}\hat{B}^{\dagger} \qquad \text{(III.3.14)}$$

allows one to get an expanded form where each term is self-adjoint. After identifying $\hat{T}$ with $\hat{T}_1(\varepsilon)$, so that $\hat{B} \equiv \hat{T}_0(a)\hat{\omega}^{\dagger}\hat{V}'$, $\hat{C} \equiv \hat{T}_0(a)\hat{P}_1$, one finds

$$\hat{T}_1(\varepsilon) = \frac{1}{2}\left[\hat{T}_0(a)\hat{P}_1 + \hat{P}_1\hat{T}_0(a)\right] + \hat{T}_0(a)\hat{\omega}^{\dagger}\hat{V}'_Q\hat{\omega}\hat{T}_0(a) +$$
$$+ \hat{T}_0(a)\hat{\omega}^{\dagger}\hat{V}'\hat{T}_1(\varepsilon)\hat{V}'\hat{\omega}\hat{T}_0(a) \qquad \text{(III.3.15)}$$

where

$$\hat{V}'_Q = \frac{1}{2}(\hat{V}'\hat{Q} + \hat{Q}\hat{V}') \qquad \text{(III.3.16)}$$

If we now multiply to the left and right by $\hat{P}_1$ and take into account that $\hat{P}_1\hat{T}_1(\varepsilon)\hat{P}_1 = \hat{T}_1(\varepsilon)$, we obtain

$$\hat{T}_1(\varepsilon) = \hat{P}_1\hat{T}_0(a)\hat{P}_1 + \hat{P}_1\hat{T}_0(a)\hat{\omega}^\dagger\hat{V}'_Q\hat{\omega}\hat{T}_0(a)\hat{P}_1 + \\ + \hat{P}_1\hat{T}_0(a)\hat{\omega}^\dagger\hat{V}'\hat{T}_1(\varepsilon)\hat{V}'\hat{\omega}\hat{T}_0(a)\hat{P}_1 \tag{III.3.17}$$

Iteration of this formula provides an expansion where each term is self-adjoint as well as contained in the subspace projected out by $\hat{P}_1$. Instead of eq. (III.3.17) one can also use the similar result

$$\hat{T}_1(\varepsilon) = \hat{P}_1\hat{T}_0(a)\hat{P}_1 + \hat{P}_1\hat{T}_0(a)\hat{\omega}^\dagger\hat{V}'\hat{Q}\hat{\omega}\hat{T}_0(a)\hat{P}_1 + \\ + \hat{P}_1\hat{T}_0(a)\hat{\omega}^\dagger\hat{V}'\hat{T}_1(\varepsilon)\hat{V}'\hat{\omega}\hat{T}_0(a)\hat{P}_1 \tag{III.3.18}$$

obtained by a simple manipulation of the second term at the right-hand side of eq. (III.3.17), exploiting the fact that $\hat{P}_1\hat{T}_0(a)\hat{\omega}^\dagger\hat{Q}\hat{V}'\hat{\omega}\hat{T}_0(a)\hat{P}_1 = \hat{P}_1\hat{T}_0(a)\hat{\omega}^\dagger\hat{V}'\hat{Q}\hat{\omega}\hat{T}_0(a)\hat{P}_1$, a consequence of the important identity

$$\left[\hat{T}_0(a),\hat{P}_1\right] = \hat{T}_0(a)\hat{\omega}^\dagger\left[\hat{Q},\hat{V}'\right]\hat{\omega}\hat{T}_0(a) \tag{III.3.19}$$

Here is the demonstration of the relation (III.3.19).
From eq. (III.3.2), after multiplying to the left by $\hat{\omega}^\dagger$ and to the right by $\hat{\omega}$, it follows

$$\hat{\omega}^\dagger\hat{H}_0\hat{P}_1 - \hat{P}_1\hat{H}_0\hat{\omega} = \hat{\omega}^\dagger[\hat{Q},\hat{V}]\hat{\omega} = \hat{\omega}^\dagger[\hat{Q},\hat{V}']\hat{\omega}$$

because $\hat{P}_1 = \hat{Q}\hat{\omega} = \hat{\omega}^\dagger\hat{Q}$. The use of eq. (III.3.5) immediately leads to

$$\hat{H}_0\hat{P}_1 - \hat{P}_1\hat{H}_0 = \hat{\omega}^\dagger[\hat{Q},\hat{V}']\hat{\omega}$$

and successively to

$$\hat{P}_1(a - \hat{H}_0) - (a - \hat{H}_0)\hat{P}_1 = \hat{\omega}^\dagger[\hat{Q},\hat{V}']\hat{\omega}$$

a being an arbitrary complex number. Hence,

$$(a - \hat{H}_0)^{-1}\hat{P}_1 - \hat{P}_1(a - \hat{H}_0)^{-1} = (a - \hat{H}_0)^{-1}\hat{\omega}^\dagger\left[\hat{Q},\hat{V}'\right]\hat{\omega}(a - \hat{H}_0)^{-1}$$

and finally

$$\left[\hat{T}_0(a),\hat{P}_1\right] = \hat{T}_0(a)\hat{\omega}^\dagger\left[\hat{Q},\hat{V}'\right]\hat{\omega}\hat{T}_0(a)$$

because, as already shown in due course, $[a - \hat{H}_0]^{-1}\hat{\omega}^\dagger = \hat{T}_0(a)\hat{\omega}^\dagger$.

The full expression for $\hat{T}_1(\varepsilon)$ corresponding to eqs. (III.3.10),

(III.3.15), (III.3.18) are to be considered entirely equivalent and, as already remarked, independent of the actual value of the parameter a. The same is by no means true for truncated expansions as well as individual terms, so that different choices for a should be expected to lead to differently effective approximations to $\hat{T}_1(\varepsilon)$.

There are two standard choices for a and two corresponding expansions for $\hat{T}_1(\varepsilon)$:

i) $a = \varepsilon$ (Brillouin-Wigner-type expansion): then $\hat{V}' = \hat{V}$;

ii) $a = E_0$ (Rayleigh-Schrödinger-type expansion), so that $\hat{V}' = \hat{V} - (\varepsilon - E_0)$.

We shall dwell on the second choice, $a = E_0$, which generates the very important Rayleigh-Schrödinger-type symmetry-adapted perturbation theory. After putting $\varepsilon_1 = \varepsilon = E$ in eq. (III.3.4), it follows

$$E = E_0 + \frac{\langle \varphi_0 | \hat{V}\hat{Q} + \hat{V}\hat{T}_1(E)\hat{V} | \varphi_0 \rangle}{\langle \varphi_0 | Q | \varphi_0 \rangle} = \qquad \text{(III.3.20)}$$
$$= E_0 + E_{01}^{(Q)} + E_{02}^{(Q)} + E_{03}^{(Q)} + \ldots..$$

where we have used $\hat{V}' = \hat{V} - (E - E_0) = \hat{V} - E_{01}^{(Q)} + E_{02}^{(Q)} + E_{03}^{(Q)} + \ldots.$, $E_{0n}^{(Q)}$ being the n-th order contribution to the change $E - E_0$ caused by the perturbation $\hat{V}$. Since different expansions for $\hat{T}_1(E)$ are available, as previously anticipated, different forms for the individual contributions $E_{0n}^{(Q)}$ are to be expected.

III.4. The Murrell-Shaw, Musher-Amos and Eisenschitz-London, Hirschfelder-Van der Avoird Formalisms.

We now propose to look into some explicit perturbation schemes which stem from eq. (III.3.20), according to the kind of expansion employed for the reduced resolvent $\hat{T}_1(E)$. We shall limit ourselves to the terms $E_{01}^{(Q)}$, $E_{02}^{(Q)}$, $E_{03}^{(Q)}$; in fact, if $E_{02}^{(Q)}$ (and still more $E_{03}^{(Q)}$) is anything but trivial to be obtained, the higher order terms appear hardly of interest at all, due to the frightfully increasing level of difficulty involved in their evaluation.

The first-order energy $E_{01}^{(Q)}$ is independent of $\hat{T}_1(E)$, so that it has the same structure whatever expansion is used:

$$E_{01}^{(Q)} = \frac{\langle\varphi_0|\hat{V}\hat{Q}|\varphi_0\rangle}{\langle\varphi_0|\hat{Q}|\varphi_0\rangle} = \frac{\langle\varphi_0|\hat{V}_Q|\varphi_0\rangle}{\langle\varphi_0|\hat{Q}|\varphi_0\rangle} \quad \text{(III.4.1)}$$

where the last equality descends from the recognition that $\langle\varphi_0|[\hat{V},\hat{Q}]|\varphi_0\rangle = 0$, an immediate consequence of eq. (III.3.2). The operator $\hat{V}_Q$

$$\hat{V}_Q = \frac{1}{2}(\hat{V}\hat{Q} + \hat{Q}\hat{V}) \quad \text{(III.4.2)}$$

is clearly the hermiticized form of the operator quantity $\hat{V}\hat{Q}$.

It is a simple matter to find another useful and important interpretation of $E_{01}^{(Q)}$. If we calculate the expectation value of the Hamiltonian operator $\hat{H}$ in the state $\hat{Q}|\varphi_0\rangle$, (clearly an approximate symmetry adapted eigenstate to the Hamiltonian operator $\hat{H} = \hat{H}_0 + \hat{V}$), the following upper bound to the lowest eigenvalue to $\hat{H}$ pertaining to the symmetry $\hat{Q}$ is obtained

$$\langle\hat{H}\rangle_{HL} = \frac{\langle\hat{Q}\varphi_0|\hat{H}|\hat{Q}\varphi_0\rangle}{\langle\hat{Q}\varphi_0|\hat{Q}\varphi_0\rangle} = \frac{\langle\varphi_0|\hat{H}\hat{Q}|\varphi_0\rangle}{\langle\varphi_0|Q|\varphi_0\rangle} = $$
$$= E_0 + \frac{\langle\varphi_0|\hat{V}\hat{Q}|\varphi_0\rangle}{\langle\varphi_0|Q|\varphi_0\rangle} \equiv E_0 + E_{01}^{(Q)} \quad \text{(III.4.3)}$$

because $\hat{H}_0|\varphi_0\rangle = E_0|\varphi_0\rangle$.

As a straightforward exemplification of the preceding remarks, let us consider the case of two hydrogen atoms with nuclei at a and b respectively, so that $|\varphi_0(1,2)\rangle = |1s_a(1)\rangle|1s_b(2)\rangle$ is the unperturbed ground state of the system. The projectors $\hat{Q}^{(+)} = \frac{1}{2}[\hat{1} + (12)]$ and $\hat{Q}^{(-)} = \frac{1}{2}[\hat{1} - (12)]$ (which clearly coincide with $\hat{\rho}^{[2]}$ and $\hat{\rho}^{[1^2]}$ respectively, see eq. (II.3.1)) give us $\hat{Q}^{(\pm)}|\varphi_0(1,2)\rangle = (2^{-1})\left[|1s_a(1)\rangle|1s_b(2)\rangle \pm |1s_b(1)\rangle|1s_a(2)\rangle\right]$, i.e. the (non-normalized) "valence-bond" or Heitler-London (approximate) eigenstates corresponding to the lowest singlet (+) and triplet (-) states, respectively [4,7,33,34]. $\langle\hat{H}\rangle_{HL}$ is commonly referred to as Heitler-London (HL) energy, while $E_{01}^{(Q)}$ is the first-order HL energy, the lowest order approximation to the interaction energy.

Second- and third-order energies $E_{02}^{(Q)}$, $E_{03}^{(Q)}$ assume a different form according to the expansion used for $\hat{T}_1(E)$. If we substitute eq.

(III.3.10) for $\hat{T}_1(E)$ into eq. (III.3.20) and expand the operator quantity $[\hat{1} - \hat{T}_0(E_0)\hat{\omega}^{\dagger}\hat{V}']^{-1}$ according to eq. (III.3.11), we get

$$E_{02}^{(Q)} = N^{-1}\langle\varphi_0|\hat{V}\hat{R}_0\hat{P}_1\hat{V}|\varphi_0\rangle \quad \text{(III.4.4)}$$

$$E_{03}^{(Q)} = N^{-1}\langle\varphi_0|\hat{V}\hat{R}_0\hat{\omega}^{\dagger}(\hat{V} - E_{01}^{(Q)})\hat{R}_0\hat{P}_1\hat{V}|\varphi_0\rangle \quad \text{(III.4.5)}$$

where we have put

$$N = \langle\varphi_0|\hat{Q}|\varphi_0\rangle \quad \text{(III.4.6)}$$

and

$$\hat{R}_0 \equiv \hat{T}_0(E_0) = [E_0 - \hat{H}_0]^{-1}\Big[1 - |\varphi_0\rangle\langle\varphi_0|\Big] \quad \text{(III.4.7)}$$

is the well known Rayleigh-Schrödinger resolvent of the ordinary perturbation theory [19,35].

Taking into account that $\hat{P}_1 = \hat{Q} - \hat{O}$ (see eq. (III.2.3)), it follows in a simple way

$$\hat{P}_1\hat{V}|\varphi_0\rangle = \hat{Q}(\hat{V} - E_{01}^{(Q)})|\varphi_0\rangle \quad \text{(III.4.8)}$$

so that $E_{02}^{(Q)}$ can be cast into the form

$$E_{02}^{(Q)} = N^{-1}\langle\varphi_0|\hat{V}\hat{R}_0\hat{Q}(\hat{V} - E_{01}^{(Q)})|\varphi_0\rangle \quad \text{(III.4.9)}$$

A manipulation of eq. (III.4.5) founded on the identity $\hat{R}_0\hat{\omega}^{\dagger} = R_0\Big[\hat{1} - N^{-1}\hat{Q}|\varphi_0\rangle\langle\varphi_0|\Big]$ allows $E_{03}^{(Q)}$ to be written in the alternative form

$$E_{03}^{(Q)} = N^{-1}\Big[\langle\varphi_0|\hat{V}\hat{R}_0(\hat{V} - E_{01}^{(Q)})\hat{R}_0\hat{Q}(\hat{V} - E_{01}^{(Q)})|\varphi_0\rangle - E_{02}^{(Q)}\langle\varphi_0|\hat{V}\hat{R}_0\hat{Q}|\varphi_0\rangle\Big] \quad \text{(III.4.10)}$$

Eqs. (III.4.9) and (III.4.10) correspond to the lowest perturbation corrections which improve the HL energy, eq. (III.4.3). They were independently obtained by Murrel, Shaw [36,37] and Musher, Amos [38,39] by using different approaches and it is common practice to refer globally to them as the second- and third-order contributions evaluated in the Murrel-Shaw, Musher-Amos (MS-MA) perturbation scheme.

If the reduced resolvent $\hat{T}_1(E)$ in eq. (III.3.20) is expanded according to the explicitly self-adjoint eq. (III.3.15), one finds symmetrized forms of the MS-MA equations [40,41].

We pass now to consider the expressions for $E_{02}^{(Q)}$ and $E_{03}^{(Q)}$ which result from eq. (III.3.20) by using the expression (III.3.18) for $\hat{T}_1(E)$. It is immediate to write down

$$
\begin{aligned}
E_{02}^{(Q)} &= N^{-1}\langle\varphi_0|\hat{V}\hat{P}_1\hat{R}_0\hat{P}_1\hat{V}|\varphi_0\rangle \\
&= N^{-1}\langle\varphi_0|(\hat{V} - E_{01}^{(Q)})\hat{Q}\hat{R}_0\hat{Q}(\hat{V} - E_{01}^{(Q)})|\varphi_0\rangle
\end{aligned}
\qquad \text{(III.4.11)}
$$

where the last equality stems directly from eq. (III.4.8). Concerning the third-order contribution,

$$E_{03}^{(Q)} = N^{-1}\langle\varphi_0|\hat{V}\hat{P}_1\hat{R}_0\hat{\omega}^{\dagger}(\hat{V} - E_{01}^{(Q)})\hat{Q}\hat{\omega}\hat{R}_0\hat{P}_1\hat{V}|\varphi_0\rangle$$

after some manipulations founded on eq. (III.4.8), the definitions of $\hat{\omega}$ and $\hat{\omega}^{\dagger}$ and eq. (III.3.2), the following equivalent form can be obtained

$$
\begin{aligned}
E_{03}^{(Q)} = N^{-1}\{&\langle\varphi_0|(\hat{V} - E_{01}^{(Q)})\hat{Q}\hat{R}_0(\hat{V} - E_{01}^{(Q)})\hat{Q}\hat{R}_0\hat{Q}(\hat{V} - E_{01}^{(Q)})|\varphi_0\rangle \\
&- E_{02}^{(Q)}[\langle\varphi_0|(\hat{V} - E_{01}^{(Q)})\hat{Q}\hat{R}_0\hat{Q}|\varphi_0\rangle + \langle\varphi_0|\hat{Q}\hat{R}_0\hat{Q}(\hat{V} - E_{01}^{(Q)})|\varphi_0\rangle]\}
\end{aligned}
\qquad \text{(III.4.12)}
$$

$E_{02}^{(Q)}$ being the 2-nd order energy, eq. (III.4.11).

Eqs. (III.4.11), (III.4.12) provide expressions for the second- and third-order corrections to the HL energy, as evaluated in the <u>Eisenschitz-London</u>, <u>Hirschfelder</u>, <u>van der Avoird</u> (EL-HAV) <u>perturbation scheme</u>* [42-45].

The comparison of the same order contributions which arise from

---

*Some historical notes about this approach do not seem out of place. The roots of the method can be traced back to the far-off paper by Eisenschitz and London (EL) [46], commonly regarded as the starting point of the quantum-mechanical treatment of atomic and molecular interactions. A compact and modern reformulation of the basic Eisenschitz and London's ideas was long after given by van der Avoird [34], who also developed a Schrödinger-type perturbation theory founded on a wave-operator formalism [42,44], quite similar to that we have presented here. At the same time, Hirschfelder presented his modification of van der Avoird's approach [43], that was shown to be equivalent to van der Avoird's treatment [45]. For these reasons, the HAV perturbation scheme is rightfully initialled EL-HAV.

different $\hat{T}_1(E)$ expansions (i.e. different perturbation schemes) shows very clearly their non-equivalence; this is already evident in the case of the second-order energy, eqs. (III.4.9) and (III.4.11). Such a behaviour is a peculiar feature of the truncated symmetry-adapted perturbative developments, where the concept of perturbation order ceases to be really meaningful, a straightforward inference from eq. (III.3.2). In order to better appreciate th statement, we establish a connection between $E_{02}^{(Q)}$(EL-HAV), the EL-HAV 2-nd order energy and $E_{02}^{(Q)}$ (MS-MA), the analogous quantity in the MSMA formalism. From eq. (III.3.17) after putting $a = E_0$ and $\hat{T}_0(E_0) \equiv \hat{R}_0$ (eq. (III.4.7)), it follows $\hat{P}_1\hat{R}_0 = \hat{R}_0\hat{P}_1 - \hat{R}_0\hat{\omega}^\dagger[Q,V]\hat{\omega}\hat{R}_0$; therefore, from eq. (III.4.11) we get

$$E_{02}^{(Q)}(\text{EL-HAV}) = E_{02}^{(Q)}(\text{MSMA}) - N^{-1}\langle\varphi_0|\hat{V}\hat{R}_0\hat{\omega}^\dagger[\hat{Q},\hat{V}]\hat{\omega}\hat{R}_0\hat{Q}(\hat{V} - E_{01}^{(Q)})|\varphi_0\rangle \tag{III.4.13}$$

i.e. $E_{02}^{(Q)}$ (EL-HAV) and $E_{02}^{(Q)}$ (MSMA) differ by a quantity that we should define of third-order.

Some other remarks concerning the 2-nd (and 3-nd) order energies in the EL-HAV an MS-MA formalism are important. If we put

$$|\varphi_{01}^{(Q)}\rangle = \hat{R}_0\hat{Q}(\hat{V} - E_{01}^{(Q)})|\varphi_0\rangle \tag{III.4.14}$$

a quantity explicitly depending on the projector $\hat{Q}$ but without any simple symmetry property, the 2-nd energy in the EL-HAV scheme can be rewritten as

$$\begin{aligned} E_{02}^{(Q)}(\text{EL-HAV}) &= N^{-1}\langle\varphi_0|(\hat{V} - E_{01}^{(Q)})\hat{Q}|\varphi_{01}^{(Q)}\rangle \\ &= N^{-1}\langle\hat{Q}(\hat{V} - E_{01}^{(Q)})\varphi_0|\hat{Q}\varphi_{01}^{(Q)}\rangle \end{aligned}$$

i.e. a scalar product between kets with manifest Q-symmetry properties. $E_{03}^{(Q)}$ (EL-HAV) can also be expressed in a form which emphasizes this feature. If we now express the 2-nd order energy in the MS-MA formalism in terms of $|\varphi_{01}^{(Q)}\rangle$, we get

$$E_{02}^{(Q)}(\text{MS-MA}) = N^{-1}\langle\varphi_0|\hat{V}|\varphi_{01}^{(Q)}\rangle$$

a scalar product between quantities which in general belong only partially to the $\hat{Q}$ - subspace of the Hilbert space [this result can be extended also to $E_{03}^{(Q)}$ (MS-MA)]. The recover at the 2-nd order of the $\hat{Q}$-symmetry by the MS-MA formalism requires 3-rd order contributions, as result from eq. (III.4.13).

As a conclusion of this section, it is appropriate to point out that in the case $[\hat{H},\hat{Q}] = [\hat{H}_0,\hat{Q}] = 0$ (i.e. the constants of motion to $\hat{H}$ are also constants of motion to $\hat{H}_0$), one also deduces $[\hat{V},\hat{Q}] = 0$, so that the above seen "third-order portion" to E_{02} vanishes exactly and the second-order results arising from different expressions of $\hat{T}_1(E)$ become identical. Such a conclusion can be generalized to higher-order contributions as well and corresponds to the case of the <u>standard Rayleigh-Schrödinger perturbation theory</u>; if we limit ourselves to consider here the case of a non-degenerate unperturbed eigenstate $|\varphi_0\rangle$, then $\hat{Q}|\varphi_0\rangle = |\varphi_0\rangle$ and we get the following unique set of perturbed energy contributions:

$$E_{01} = \langle\varphi_0|\hat{V}|\varphi_0\rangle$$

$$E_{02} = \langle\varphi_0|\hat{V}\hat{R}_0\hat{V}|\varphi_0\rangle \qquad \text{(III.4.15)}$$

$$E_{03} = \langle\varphi_0|\hat{V}\hat{R}_0(\hat{V} - E_{01})\hat{R}_0\hat{V}|\varphi_0\rangle$$

....

The abolition of the superscript (Q) from E_{0n} should be noted: it reflects the fact that now, starting from $|\varphi_0\rangle$, no symmetry adaptation has been introduced.

E_{02} and E_{03} are seen to be individuated by the ket

$$|\varphi_{01}\rangle = \hat{R}_0\hat{V}|\varphi_0\rangle \qquad \text{(III.4.16)}$$

It is interesting to note that $E_{02}^{(Q)}$ (MS-MA) (but not $E_{03}^{(Q)}$ (MS-MA)) can be obtained starting from $|\varphi_{01}\rangle$ instead of $|\varphi_{01}^{(Q)}\rangle$, eq. (III.4.14). In fact,

$$E_{02}^{(Q)}\text{(MS-MA)} = N^{-1}\langle\varphi_{01}|\hat{Q}(\hat{V} - E_{01}^{(Q)})\varphi_0\rangle. \qquad \text{(III.4.16')}$$

This latter result is of value for applications, as we shall see in

Chap. V.

III.5. Multidimensional Partitioning Technique and Degenerate Perturbation Theory.

As well known, the extension of the ordinary perturbation theory to problems involving degenerate levels is generally not painless. From this angle, the partitioning technique cannot work wonders, but it surely allows a compact and elegant presentation of the subject [32], the necessity of a multidimensional reference space (see sec. III.2) being the main toll to be paid for setting up the proper generalization.

In order to get directly to the heart of the matter, we shall consider the case where the standard (i.e. non-symmetry-adapted) perturbation theory is applied to a system described by a Hamiltonian operator that can be partitioned in the form (III.3.1), i.e. $\hat{H} = \hat{H}_0 + \hat{V}$, and pay our attention to the unperturbed level E_0, supposed to be p times degenerate, so that

$$\hat{H}_0 |\varphi_{ok}\rangle = E_0 |\varphi_{ok}\rangle, \quad k = 1,2,..,p \qquad \text{(III.5.1)}$$

The unperturbed eigenvectors $|\varphi_{ok}\rangle$ belonging to E_0 will be assumed to form an orthonormal set, i.e. $\langle\varphi_{ok}|\varphi_{oj}\rangle = \delta_{kj}$.

As a reference space we shall now take a (linear) manifold spanned by the p eigenvectors $|\varphi_{01}\rangle,\ldots,|\varphi_{op}\rangle$ to $\hat{H}_0$. The associated projection operator

$$\hat{O} = \sum_{k=1}^{p} |\varphi_{ok}\rangle\langle\varphi_{ok}| \equiv |\boldsymbol{\varphi}_o\rangle\langle\boldsymbol{\varphi}_o| \qquad \text{(III.5.2)}$$

clearly satisfies $\hat{O}^2 = \hat{O}$, $\hat{O}^\dagger = \hat{O}$, $\mathrm{Tr}\hat{O} = p$, $\hat{O}|\boldsymbol{\varphi}_o\rangle = |\boldsymbol{\varphi}_o\rangle$[we have introduced the notation $|\boldsymbol{\varphi}_o\rangle$ for the row-matrix $(|\varphi_{01}\rangle\ |\varphi_{01}\rangle\ldots|\varphi_{op}\rangle)$].

The operator for the complementary space

$$\hat{P} = \hat{1} - \hat{O} \qquad \text{(III.5.3)}$$

is also a projector; obviously it is such that $\hat{P}^2 = \hat{P}$, $\hat{P}^\dagger = \hat{P}$, $\hat{P}\hat{O} = \hat{O}\hat{P} = 0$, $\hat{P}|\boldsymbol{\varphi}_o\rangle = 0$.

Proceeding in complete analogy with sec. III.2, we now introduce the auxiliary Hamiltonian operator $\hat{H} = \hat{P}\hat{H}\hat{P}$ and its reduced resolvent $\hat{T}(\varepsilon) = [\varepsilon - \hat{H}]^{-1}\hat{P}$, which satisfies the following fundamental relations:

$$\hat{P}(\varepsilon - \hat{H})\hat{T}(\varepsilon) = \hat{P}, \quad \hat{P}\hat{T}(\varepsilon) = \hat{T}(\varepsilon)\hat{P} = \hat{T}(\varepsilon), \quad \hat{T}^{\dagger}(\varepsilon) = \hat{T}(\varepsilon) \tag{III.5.4}$$

The study of the eigenvalue problem which we are interested in,

$$\hat{H}|\Psi(E)\rangle = (\hat{H}_0 + \hat{V})|\Psi(E)\rangle = E|\Psi(E)\rangle \tag{III.5.5}$$

will be pursued by considering the "trial eigenvector" $|\Psi(\varepsilon)\rangle$

$$|\Psi(\varepsilon)\rangle \quad [\hat{1} + \hat{T}(\varepsilon)\hat{H}]|\Phi\rangle \tag{III.5.6}$$

where $|\Phi\rangle = |\boldsymbol{\varphi}_0\rangle\mathbb{C}$ and $\mathbb{C}$ is presently an arbitrary column vector. $|\Phi\rangle$ clearly lies in the reference space, so that $\hat{P}|\Phi\rangle = 0$ and, therefore, $\hat{P}(\hat{H} - \varepsilon)|\Psi(\varepsilon)\rangle = 0$. As in sec. III.2, we then find

$$(\varepsilon - \hat{H})|\Psi(\varepsilon)\rangle = \hat{O}(\varepsilon - \hat{H})|\Psi(\varepsilon)\rangle = |\boldsymbol{\varphi}_0\rangle\langle\boldsymbol{\varphi}_0|\varepsilon - \hat{H}|\Psi(\varepsilon)\rangle$$

i.e.

$$(\hat{H} - \varepsilon)|\Psi(\varepsilon)\rangle = |\boldsymbol{\varphi}_0\rangle\{\langle\boldsymbol{\varphi}_0|\hat{H} + \hat{H}\hat{T}(\varepsilon)\hat{H}|\boldsymbol{\varphi}_0\rangle - \varepsilon\mathbb{1}\}\mathbb{C} \tag{III.5.7}$$

where $\mathbb{1}$ is the (p×p) identity matrix.

If we now choose the column vector $\mathbb{C}$ in such a way that

$$\langle\boldsymbol{\varphi}_0|\hat{H} + \hat{H}\hat{T}(\varepsilon)\hat{H}|\boldsymbol{\varphi}_0\rangle\,\mathbb{C} = \bar{E}\,\mathbb{C} \tag{III.5.8}$$

the condition for non-trivial solutions $\mathbb{C}$ requires the following secular equation to be satisfied:

$$\|\langle\boldsymbol{\varphi}_0|\hat{H} + \hat{H}\hat{T}(\varepsilon)\,\hat{H} - \bar{E}\mathbb{1}|\boldsymbol{\varphi}_0\rangle\| \equiv \det|\langle\boldsymbol{\varphi}_0|\hat{H} + \hat{H}\hat{T}(\varepsilon)\hat{H} - \bar{E}\mathbb{1}|\boldsymbol{\varphi}_0\rangle| = 0. \tag{III.5.9}$$

The inhomogeneous equation (III.5.7) therefore becomes

$$(\hat{H} - \varepsilon)|\Psi(\varepsilon)\rangle = (\bar{E} - \varepsilon)|\Phi\rangle \tag{III.5.10}$$

so that a solution to the Schrödinger equation (III.5.5) is found whenever $\overline{E}(\varepsilon) = \varepsilon = E$. The secular equation (III.5.9) defines a p-fold function $\overline{E}(\varepsilon)$: each branch of this function has the property that, between ε and $\overline{E}(\varepsilon)$, there is at least one true eigenvalue E satisfying $E = \overline{E}(E)$ ("bracketing theorem"), just as in the case of the one-dimensional reference space (see sec. III.2).

If the partition of the Hamiltonian operator $H = H_0 + V$ is exploited in eq. (III.5.9), the following result is easily derived

$$\| \langle \varphi_0 | E_0 \mathbb{1} + \hat{V} + \hat{V}\hat{T}(\varepsilon)\hat{V} - \overline{E}\mathbb{1} | \varphi_0 \rangle \| = 0 \qquad \text{(III.5.11)}$$

Important perturbative expansions descend from the iteration of

$$\hat{T}(\varepsilon) = \hat{T}_0(a) + \hat{T}_0(a)\hat{V}'\hat{T}(\varepsilon) \qquad \text{(III.5.12)}$$

a relation readily got by using manipulations analogous to those used in the previous section, with $\hat{V}'$ and $\hat{T}_0(a)$ given by eqs. (III.3.8) and (III.3.9), respectively.

The choice a = E defines the Brillouin-Wigner unperturbed resolvent $\hat{T}_0(E) \equiv \hat{T}_0$ and the corresponding expansion results

$$\| \langle \varphi_0 | (E_0 - E)\mathbb{1} + \sum_{k=1}^{\infty} \hat{V}(\hat{T}_0\hat{V})^{k-1} | \varphi_0 \rangle \| = 0 \qquad \text{(III.5.13)}$$

The other important choice, $a = E_0$, which defines the Rayleigh-Schrödinger unperturbed resolvent $\hat{T}_0(E_0) \equiv \hat{R}_0$, leads to a secular equation of the form

$$\| \langle \varphi_0 | (E_0 - E)\mathbb{1} + \hat{V} + \hat{V}\hat{R}_0\hat{V} + \dots | \varphi_0 \rangle \| = 0 \qquad \text{(III.5.14)}$$

In actual applications, a truncation of the sums after at most the first few terms will be the standard occurrence.

Eq. (III.5.14) is a form of the so-called Van Vleck's degenerate perturbation theory [47]. Such formalism has been remarked to have advantages over the conventional degenerate Rayleigh-Schrödinger perturbation theory [48], essentially because no removal of degeneracy is necessary at the outset, a procedure which is hard when the degeneracy is not lifted at the first-order.

Symmetry-adapted, degenerate perturbation theory will be not considered here; for an operational formalism, along lines similar to those developed by us in the non-degenerate context, one can profitably turn to ref. [49].

III.6. Upper and Lower Bounds in Second-Order Perturbation Theory by Inner Projection Technique.

Since many physical effects and properties are described in terms of second-order perturbation theory, whose actual evaluation generally requires the introduction of approximations, techniques which provide upper and lower bounds for these quantities are indeed welcome for a better appreciation of the theoretical results.

The proper mathematical tools are operator inequalities [40,50-52], which can be introduced by making reference to a linear manifold M_n, spanned by the n linearly independent vectors $\{|f_1\rangle, |f_2\rangle, \ldots |f_n\rangle\} \equiv |\mathbf{f}\rangle$. The projection operator associated with M_n is

$$\hat{O} = |\mathbf{f}\rangle \boldsymbol{\Delta}^{-1} \langle \mathbf{f}| \equiv \sum_{jk}^{n} |f_j\rangle (\boldsymbol{\Delta}^{-1})_{jk} \langle f_k|$$

where $\boldsymbol{\Delta} = \langle \mathbf{f}|\mathbf{f}\rangle$ is the metric matrix. Clearly $\hat{O}^2 = \hat{O}$, $\hat{O}^\dagger = \hat{O}$, $\mathrm{Tr}\hat{O} = n$; moreover, from $\hat{O} = \hat{O}^\dagger\hat{O} \geq 0$ and $\hat{1} - \hat{O} = (\hat{1} - \hat{O})^\dagger(\hat{1} - \hat{O}) \geq 0$, it follows

$$0 \leq \hat{O} \leq 1 \tag{III.6.1}$$

a result valid for any projector. If now $\hat{A}$ denotes a positive definite operator and $\hat{A}^{\frac{1}{2}}$ its "square root", from eq. (III.6.1) one immediately gets

$$0 \leq \hat{A}^{\frac{1}{2}}\hat{O}\hat{A}^{\frac{1}{2}} \leq \hat{A} \tag{III.6.2}$$

The operator $\hat{A}' = \hat{A}^{\frac{1}{2}}\hat{O}\hat{A}^{\frac{1}{2}}$ is called the inner projection of $\hat{A}$ on the manifold M_n; we note that $0 \leq \hat{A}' \leq \hat{A}$ and $\hat{A}'$ approaches $\hat{A}$ as the set $|\mathbf{f}\rangle$ becomes complete.

In addition to the manifold spanned by $|\mathbf{f}\rangle$, we need another manifold spanned by $\{|h_1\rangle, \ldots, |h_n\rangle\} \equiv |\mathbf{h}\rangle$, where

$$|\mathbf{f}\rangle = \hat{A}^{-\frac{1}{2}} |\mathbf{h}\rangle \tag{III.6.3}$$

We therefore obtain the following alternatives for the inner projection operator $\hat{A}'$:

$$\hat{A}' = \hat{A}^{\frac{1}{2}}|\mathbf{f}\rangle\langle\mathbf{f}|\mathbf{f}\rangle^{-1}\langle\mathbf{f}|\hat{A}^{\frac{1}{2}} = |\mathbf{h}\rangle\langle\mathbf{h}|\hat{A}^{-1}|\mathbf{h}\rangle^{-1}\langle\mathbf{h}| \qquad \text{(III.6.4)}$$

For a <u>negative</u> <u>definite</u> operator $\hat{B}$, one obtains by putting $\hat{A} = -\hat{B}^{-1}$ in the last form of eq. (III.6.4):

$$|\mathbf{h}\rangle\langle\mathbf{h}|\hat{B}|\mathbf{h}\rangle^{-1}\langle\mathbf{h}| \geq \hat{B}^{-1}$$

and then

$$\langle\tilde{\varphi}|\mathbf{h}\rangle\langle\mathbf{h}|\hat{B}|\mathbf{h}\rangle^{-1}\langle\mathbf{h}|\tilde{\varphi}\rangle \geq \langle\tilde{\varphi}|\hat{B}^{-1}|\tilde{\varphi}\rangle \qquad \text{(III.6.5)}$$

$|\tilde{\varphi}\rangle$ being an arbitrary vector in the domain of $\hat{A}$, for which $\langle\tilde{\varphi}|\mathbf{h}\rangle$ exists.

Let us now consider the unperturbed reduced resolvent $\hat{R}_0$ defined by eq. (III.4.20), $\hat{R}_0 = [E_0 - \hat{H}_0]^{-1}[\hat{1} - |\varphi_0\rangle\langle\varphi_0|]$, where $\hat{H}_0|\varphi_n\rangle = E_n|\varphi_n\rangle$. From the spectral resolution of $\hat{R}_0$,

$$\hat{R}_0 = \sum_{n\neq 0} \frac{|\varphi_n\rangle\langle\varphi_n|}{E_0 - E_n}, \qquad \text{(III.6.6)}$$

it easily descends that $\hat{R}_0 \leq 0$, so that the disequality (III.6.5) is valid for $\hat{B}^{-1} \equiv \hat{R}_0$. Hence,

$$\langle\tilde{\varphi}|\mathbf{h}\rangle\langle\mathbf{h}|E_0 - H_0|\mathbf{h}\rangle^{-1}\langle\mathbf{h}|\tilde{\varphi}\rangle \geq \langle\tilde{\varphi}|\hat{R}_0|\tilde{\varphi}\rangle \qquad \text{(III.6.7)}$$

provided that the set $|\mathbf{h}\rangle \equiv \{|h_1\rangle,\ldots,|h_n\rangle\}$, otherwise arbitrary, satisfies $\langle\varphi_0|\mathbf{h}\rangle = 0$.

If we now choose $|\tilde{\varphi}\rangle = \langle\varphi_0|\hat{Q}|\varphi_0\rangle^{-\frac{1}{2}}\hat{P}_1\hat{V}|\varphi_0\rangle$, where $\hat{P}_1$ is the projector extensively used in the previous sections and defined by eq. (III.2.3), we get [40]

$$E_{02}^{(Q)} \leq \langle\varphi_0|\hat{Q}|\varphi_0\rangle^{-1}\langle\varphi_0|\hat{V}\hat{P}_1|\mathbf{h}\rangle\langle\mathbf{h}|E_0 - \hat{H}_0|\mathbf{h}\rangle^{-1}\langle\mathbf{h}|\hat{P}_1\hat{V}|\varphi_0\rangle \qquad \text{(III.6.8)}$$

that provides an upper bound to the EL-HAV second-order energy obtained in sec. III.4, eq. (III.4.11). Such a bound is an optimal one, in the sense that it is the best upper bound which can be obtained from the basis $|\mathbf{h}\rangle$ [50].

The problem of finding a lower bound to E_{02} is approached by considering the "remainder operator"

$$\hat{r} = \hat{R}_0 - (E_0 - E_1)^{-1}\hat{P} \tag{III.6.9}$$

where E_1 is the first excited eigenvalue to $\hat{H}_0$ and $\hat{P} = \hat{1} - |\varphi_0\rangle\langle\varphi_0|$. Clearly, $\hat{R}_0 \geq (E_0 - E_1)^{-1}\hat{P}$, so that $\hat{r} \geq 0$, i.e. the "remainder operator" is <u>positive</u> <u>definite</u>. This operator can be rewritten in the form

$$\begin{aligned} \hat{r} &= \hat{P}(E_0 - \hat{H}_0)^{-1} - \hat{P}(E_0 - E_1)^{-1} = \\ &= \hat{P}(E_1 - E_0)^{-1}(\hat{H}_0 - E_1)(\hat{H}_0 - E_0)^{-1} \end{aligned} \tag{III.6.10}$$

that puts in evidence how $\hat{r}$ has the eigenvalue zero associated with the eigenvector $|\varphi_1\rangle$ to $\hat{H}_0$. If $|\mathbb{h}\rangle$ denotes a subset $\{|h_1\rangle,\ldots,|h_n\rangle\}$ lying in the orthogonal complement to $\{|\varphi_0\rangle,|\varphi_1\rangle\}$, from eq. (III.6.4) we get

$$\hat{r} \geq \hat{r}' = (E_1 - E_0)^{-1}|\mathbb{h}\rangle\langle\mathbb{h}|(\hat{H}_0 - E_0)(\hat{H}_0 - E_1)^{-1}|\mathbb{h}\rangle^{-1}\langle\mathbb{h}|$$

Putting $(\hat{H}_0 - E_1)|\mathbb{k}\rangle = |\mathbb{h}\rangle$, it follows

$$\begin{aligned} \hat{r} \geq \hat{r}' = (E_1 - E_0)^{-1}(\hat{H}_0 - E_1)|\mathbb{k}\rangle\langle\mathbb{k}|(\hat{H}_0 - E_1)(\hat{H}_0 - E_0)|\mathbb{k}\rangle^{-1} \\ \cdot\langle\mathbb{k}|(\hat{H}_0 - E_1) \end{aligned}$$

If we, as previously, choose $|\tilde{\varphi}\rangle = \langle\varphi_0|\hat{Q}|\varphi_0\rangle^{-\frac{1}{2}}\hat{P}_1\hat{V}|\varphi_0\rangle$, there results a lower bound to the EL-HAV second-order energy, eq. (III.4.11):

$$\begin{aligned} E_{02}^{(Q)} \geq \frac{(E_0 - E_1)^{-1}}{\langle\varphi_0|\hat{Q}|\varphi_0\rangle}\Big[\langle\varphi_0|\hat{V}\hat{P}_1\hat{P}\hat{P}_1\hat{V}|\varphi_0\rangle - \langle\varphi_0|\hat{V}\hat{P}_1(\hat{H}_0 - E_1)|\mathbb{k}\rangle \\ \cdot\langle\mathbb{k}|(\hat{H}_0 - E_1)(\hat{H}_0 - E_0)|\mathbb{k}\rangle^{-1}\langle\mathbb{k}|(\hat{H}_0 - E_1)\hat{P}_1\hat{V}|\varphi_o\rangle\Big] \end{aligned} \tag{III.6.11}$$

Upper and lower bounds for the MSMA second-order energy can be obtained [40] and related to the analogous bounds for the EL-HAV energy, but we shall not get involved in the derivation.

The standard Rayleigh-Schrödinger perturbation theory is recovered as $[\hat{H},\hat{Q}] = [\hat{H}_0,\hat{Q}] = 0$, so that $[\hat{V},\hat{Q}] = 0$; in such a case simplified expressions for both upper and lower bounds to $E_{02}^{(Q)}$ stem from eq. (III.6.8) and (III.6.11) by straightforward manipulations founded on $\hat{Q} \to \hat{1}$ (case of a non-degenerate state) and $\hat{P}_1 \to \hat{1} - |\varphi_0\rangle\langle\varphi_0| = \hat{P}$.

As a conclusion of the present section, we simply limit ourselves to state that upper and lower bounds can rather directly established for the second-order energy relating to an unperturbed degenerate excited state, where both first-order energy and off-diagonal matrix elements in the secular equation vanish [53,54]. Such a situation is, for instance, encountered in the study of the dispersion energy between two atoms, one of which is in an excited degenerate state and the other in its ground state, when the interaction potential is approximated by its multipolar expansion.

IV. WHY SYMMETRY-ADAPTED PERTURBATION THEORIES ARE NEEDED?

IV.1. "Polarization Approximation": Some Hints to its Possible Inadequacy for Intermolecular Interactions.

Now that we have presented some symmetry-adapted perturbation schemes, we would like to understand the reasons that suggest or motivate their introduction for the ab-initio prediction of the interaction energy $\Delta E = E^{(ABC...)} - (E^A + E^B + E^C + ...)$ between a number of atomic or molecular partners A, B, C, ..., in a given configuration.

At first sight, in fact, the standard Rayleigh-Schrödinger perturbation method would stand as a logical candidate, the unperturbed Hamiltonian operator $\hat{H}_0$ being the sum of Hamiltonian operators for the separate subsystems A, B, C, ..., i.e. $\hat{H}_0 = \hat{H}_0^A + \hat{H}_0^B + ...$ and the perturbation Hamiltonian $\hat{V} = \hat{H} - \hat{H}_0$, with $\hat{H}$ Hamiltonian operator for the complex ABC... This very natural procedure, which allots definite electrons to each subsystem A, B, ... has been named "polarization approximation"(PA) by Hirschfelder [5,6,55]. We shall make reference to this approach by using such denomination, without disdaining however the appellation of standard Rayleigh-Schrödinger perturbation theory (RSPT).

It is very important to emphasize how such procedure of allotment, implicit in the definition of $\hat{H}_0$, causes $\hat{H}_0$ to be not symmetric under all possible exchanges of electron labels among the partners A, B, C, ..., the invariance being conserved only under the subset of exchanges within each monomer. In more technical language, if the Hamiltonian operator $\hat{H}$ for the composite system (ABC...) involves $N = N_A + N_B + N_C$ +.. electrons, N_K being the number of electrons associated with the K-th monomer, we say that $\hat{H}$ admits as a symmetry group the symmetric group S_N of N particles, while $\hat{H}_0 = \hat{H}^A(1...N_A) + \hat{H}^B(N_A+1...N_A+N_B) +$ + ..., the unperturbed Hamiltonian, is invariant only under $S_{N_A} \otimes S_{N_B} \otimes ..$, the S_N subgroup which is the direct product of the symmetric groups S_{N_K} of N_K particles.

For example, in the very simple case of two interacting helium atoms, the (electronic) Hamiltonian operator results $\hat{H}(1,2,3,4) = \sum_{i=1}^{4}[\hat{T}(i) + \hat{V}_A(i) + \hat{V}_B(i)] + \sum_{i<j}^{4}\hat{V}(i,j)$ ($\hat{T}(i) \equiv$ kinetic energy operator for

the electron i; $\hat{V}_A(i)$ ≡ potential energy operator between electron i and nucleus A; $\hat{V}(i,j)$ ≡ interelectronic potential energy operator). A possible choice for $\hat{H}_0$ could be $\hat{H}_0(1,2,3,4) \equiv \hat{H}_0^A(1,2) + \hat{H}_0^B(3,4) = \{\sum_{i=1}^{2}[\hat{T}(i) + \hat{V}^A(i)] + \hat{V}(1,2)\} + \{\sum_{i=3}^{4}[\hat{T}(i) + \hat{V}^B(i)] + \hat{V}(3,4)\}$, which allots the electrons 1 and 2 to the atom A and the electrons 3, 4 to the atom B. One immediately verifies that, unlike $\hat{H}(1,2,3,4)$, $\hat{H}_0(1,2,3,4)$ is symmetric under the exchanges $1\leftrightarrows 2$, $3\leftrightarrows 4$, but not $1\leftrightarrows 3$, $2\leftrightarrows 4$ etc.

We are, therefore, faced with the not usual situation in the perturbative context that, at least from the point of view of the <u>permutational</u> symmetry (i.e. symmetry under exchange of electrons), the unperturbed Hamiltonian $\hat{H}_0$ has a <u>lower</u> symmetry than the perturbed Hamiltonian $\hat{H}$. This occurrence has raised arguments against the possibility of using in a straightforward way the standard RSPT; here we limit ourselves to sketch two (not independent) arguments advanced against RSPT.

a) (Musher [56], Murrell - Shaw [36]).

Any perturbed solution $|\Psi\rangle$ to the Schrödinger equation corresponding to two fully interacting subsystems A, B is built up in the Hilbert space spanned by the eigenstates to the unperturbed Hamiltonian $\hat{H}_0 = \hat{H}_0^A + \hat{H}_0^B$, which are necessarily simple products of the form $|\varphi_n^A\rangle|\varphi_m^B\rangle$. $|\varphi_n^A\rangle$ ($|\varphi_m^B\rangle$) is an exact eigenstate to $\hat{H}_0^A$ ($\hat{H}_0^B$) fully antisymmetric under electron exchange (according to the Pauli principle): $|\varphi_n^A\rangle|\varphi_m^B\rangle$ is therefore not antisymmetric under exchanges of electrons between A and B.

Since the Hamiltonian operators $\hat{H}$, $\hat{H}_0$ are spin-independent (we shall always neglect the finer effects arising from the inclusion of spin-dependent terms), the perturbation $\hat{V} = \hat{H} - \hat{H}_0$ can mix with the unperturbed ground state $|\varphi_0^A\rangle|\varphi_0^B\rangle$ only those states $|\varphi_n^A\rangle|\varphi_m^B\rangle$ where $|\varphi_n^A\rangle$ has the same spin as $|\varphi_0^A\rangle$ and $|\varphi_m^B\rangle$ the same spin as $|\varphi_0^B\rangle$. $|\Psi\rangle$ will, therefore, possess the same spin as $|\varphi_0^A\rangle|\varphi_0^B\rangle$ and will <u>not</u> be totally antisymmetric (such a behaviour requires eigenstates $|\varphi_n^A\rangle$ and $|\varphi_m^B\rangle$ with all possible spins, only subjected to the constraint that the resultant spin is that of $|\Psi\rangle$).

b) (Hirschfelder [55], Certain - Bruch [5]).

Let $G \equiv \{g_i\}$ be a group of operators commuting with the Hamiltonian

$\hat{H}$, so that $[\hat{g}_i,\hat{H}] = 0$. After putting $\hat{H} = \hat{H}_0 + \hat{V}$, it follows [see eq. (III.3.2)],

$$[\hat{g}_i,\hat{H}_0] = [\hat{V},\hat{g}_i]$$

We therefore conclude that if $\hat{g}_i$ commutes with $\hat{H}_0$, it also commutes with $\hat{V}$. If however $\hat{g}_i$ does not commute with $\hat{H}_0$, a consistent perturbation procedure looks quite hard to set up, because a zero-th order quantity like $[\hat{g}_i,\hat{H}_0]$ is equal to a first order quantity $[\hat{V},\hat{g}_i]$, so making the perturbation order not uniquely defined*.
To appreciate this intriguing difficulty in a different way, let us consider the simple example of two interacting hydrogen atoms. The standard RSPT approach to the interaction energy requires a zero-th order Hamiltonian, which could be $\hat{H}_0(1,2) = [\hat{T}(1) + \hat{V}^A(1)] + [\hat{T}(2)+\hat{V}^B(2)]$, so that the perturbation $\hat{V}(1,2) = \hat{V}^B(1) + \hat{V}^A(2) + \hat{V}(1,2)$.
If $(\hat{12})$ denotes the (unitary) permutation operator which exchanges the electrons 1, 2, the transformed operator $\hat{H}_0'(1,2) = (\hat{12})\hat{H}_0(1,2)\cdot(\hat{12})^{-1} = [\hat{T}(2) + \hat{V}^A(2)] + [\hat{T}(1) + \hat{V}^B(1)] \neq \hat{H}_0(1,2)$ and, analogously, $\hat{V}'(1,2) = (\hat{12})\hat{V}(1,2)(\hat{12})^{-1} \neq \hat{V}(1,2)$, but $\hat{H}'(1,2) = \hat{H}_0'(1,2) + \hat{V}'(1,2) = \hat{H}(1,2)$, a logical consequence of $[(\hat{12}),\hat{H}(1,2)] = 0$.
If we now introduce, in the standard way, a perturbation parameter λ, so that the perturbed Hamiltonian becomes $\hat{H}(1,2;\lambda) = \hat{H}_0(1,2) + \lambda\hat{V}(1,2)$, we recognize that $\hat{H}'(1,2;\lambda) = (\hat{12})\hat{H}(1,2;\lambda)(\hat{12})^{-1} = \hat{H}_0'(1,2) + \lambda\hat{V}'(1,2)$ only equals $\hat{H}(1,2;\lambda) = \hat{H}_0(1,2) + \lambda\hat{V}(1,2)$ as $\lambda = 1$. It follows that the perturbed solutions cannot be generated by continuously varying λ from zero to unity.

These two related arguments, however, do not give, in our opinion, an entirely transparent and convincing evidence for the inadequacy of the standard RSPT when applied to intermolecular energy problems, although they touch, without a shadow of doubt, difficulties to be encountered in this field. A better and deeper appreciation of the possible reasons for this inadequacy is needed and we believe the clear analysis

*$\hat{g}_i$ actually commutes with the kinetic energy operators contained in $\hat{H}_0$; therefore $[\hat{V},\hat{g}_i]$ equals the commutator of $\hat{g}_i$ with a part of $\hat{H}_0$ [57].

put forward by Claverie [6,17] to be particularly important on this subject. In the next sections we propose to review such an analysis and to find answers for the following questions:

1) In the "polarization approximation" the appropriate Hamiltonian operator is $\hat{H}(\lambda) = \hat{H}_0 + \lambda\hat{V}$. Starting from some eigenvalue to $\hat{H}_0 \equiv \hat{H}(0)$ we are interested in some eigenvalue to $\hat{H} \equiv \hat{H}(1) = \hat{H}_0 + \hat{V}$, to be reached by Taylor expansion in the λ parameter around the point $\lambda = 0$. What is, then, the connection between the eigenvalues to $\hat{H}$ and those to $\hat{H}_0$? Such a question is very relevant, because we may only reach a given eigenvalue to $\hat{H}$, starting from a given eigenvalue to $\hat{H}_0$, if there exists a curve $E(\lambda)$ connecting these two eigenvalues $E(1)$, $E(0)$.
2) Is the convergence radius of the Taylor expansion around $\lambda = 0$ larger or smaller than unity?

We shall start by discussing the second point.

IV.2. Is the "Polarization Approximation" a Convergent Perturbative Procedure?

Let us consider the RSPT expansion for the energy

$$E(\lambda) = E_0 + \lambda E_{01} + \lambda^2 E_{02} + \ldots \qquad \text{(IV.2.1)}$$

the value of interest for the coupling parameter being $\lambda = 1$. Resorting to basic statements like Rellich theorem and Kato corollary [18,19] does not settle the matter of the convergence of eq. (IV.2.1), because these two mathematical tools give comforting assurances only for small λ values. The case of interest, $\lambda = 1$, does not seem susceptible of a general analysis, so one is forced to handle a given problem by itself or, more frequently, to accept "bona fide" the validity of the expansion.

The convergence radius R_C of the expansion (IV.2.1) is, in principle, determined by the knowledge of the coefficients E_{0n} for large n: an important investigation along this line, for the case of the interaction $H....H^+$, will be discussed in a short time, but it is right to emphasize from now the prohibitive difficulties to be encountered as one tries to pass to more complicate molecules. A second approach to R_C involves the search for the singularity λ_0 in the energy $E(\lambda)$ nearest $\lambda = 0$, because $R_C = |\lambda_0|$ [58]: also in this case, generally insurmountable difficulties are to be expected for actual molecules, so the analysis has been applied to soluble model problems, from which one hopes to gain some insight into the nature of the singularities in realistic problems.

We shall now examine from the latter point of view the one-dimensional delta-function model for one-electron diatoms, a problem rather extensively studied [59-62]. The Schrödinger equation to be solved (atomic units are used throughout)

$$\left[-\frac{1}{2}\frac{d^2}{dx^2} - Z_a\,\delta\left(x - \frac{R}{2}\right) - Z_b\,\delta\left(x + \frac{R}{2}\right)\right]\Psi_E(x) = E\Psi_E(x)$$

admits only two bound states [63], whose eigenvalues correspond to the roots of the trascendental equation

$$(Z_a - \sqrt{-2E})(Z_b - \sqrt{-2E}) = Z_a Z_b \exp(-2R\sqrt{-2E}) \qquad \text{(IV.2.2)}$$

We shall assume as unperturbed system the larger of the two atoms, say A, so that the delta potential at B acts as perturbation. After defining the following parameters

$$\lambda = Z_b/Z_a, \quad a = 2RZ_a, \quad \gamma = \sqrt{-2E}/Z_a$$

eq. (IV.2.2) can be rewritten as

$$\gamma = 1 + \lambda\left[1 + (e^{-a\gamma} - 1)/\gamma\right] \qquad \text{(IV.2.3)}$$

The expansion of $\gamma(\lambda)$ in (direct) powers of λ then provides the RSPT series for the energy $E = -\frac{1}{2}Z_a^2\gamma^2(\lambda)$, which can be shown to have the same convergence radius as the series for γ [60].

The properties of the implicit equation (IV.2.3) can be studied by considering the more general equation $z = 1 + \lambda\phi(z)$, where $\phi(z)$ is an analytic function of the complex variable z. A sufficient condition such that the remainder $R_N(\lambda)$ for the $z(\lambda)$ expansion

$$z(\lambda) = 1 + \lambda z_1 + \lambda^2 z_2 + \ldots + \lambda^N z_N + R_N(\lambda) \qquad \text{(IV.2.4)}$$

becomes vanishingly small as N increases more and more (i.e. the expansion converges), is that λ is sufficiently small so that $|z - 1| > |\lambda\phi(z)|$ on a circle in the complex plane [58,60,41]. After putting $z - 1 = r\,e^{i\theta}$, the previous disequality provides in a straightforward way the following result

$$R_{\ell b} = \max_{r>0}\left[\min_{0\leq\theta\leq 2\pi}\left(\frac{r}{|\phi(1+re^{i\theta})|}\right)\right] \qquad \text{(IV.2.5)}$$

for the lower bound $R_{\ell b}$ to the largest $|\lambda|$ for which the convergence of the series (IV.2.4) is warranted. The condition expressed by eq. (IV.2.5) corresponds to the search for the saddle point with the largest $|\lambda|$ value on the $|\lambda(z)|$ surface, a (necessary) condition for its existence at $z = 1 + r\,e^{i\theta}$ being

$$\frac{\partial|\lambda(z)|}{\partial\theta} = 0 \;, \qquad \frac{\partial|\lambda(z)|}{\partial r} = 0$$

If one puts $\lambda = (z - 1)/\phi(z)$ and $\phi(z) = u + i\,v$ (with u and v real functions), the preceding condition can be expressed in the form

$$u\,\frac{\partial u}{\partial\theta} + v\,\frac{\partial v}{\partial\theta} = 0$$

$$u(u - r\,\frac{\partial u}{\partial r}) + v(v - r\,\frac{\partial v}{\partial r}) = 0$$

and then, after exploiting the Riemann-Cauchy relations,

$$\begin{aligned} u - r\,\frac{\partial u}{\partial r} &= 0 \\ v - r\,\frac{\partial v}{\partial r} &= 0 \end{aligned} \qquad \text{(IV.2.6)}$$

The necessary conditions (IV.2.6) for a saddle point on the $|\lambda(z)|$ surface are, on the other hand, the real and imaginary components of the equation

$$\frac{d\lambda}{dz} = 0 \qquad \text{(IV.2.7)}$$

which is a (sufficient) condition for a branch point of $z(\lambda)$, i.e. a point where the branches of $z(\lambda)$ becomes equal [61]. The singularity which controls the convergence radius of the series (IV.2.4) is therefore the point where the branches of $z(\lambda)$ coalesce.

Coming back to our actual problem, $\phi(\gamma) = 1 + (e^{-a\gamma} - 1)/\gamma$, and the branch point is located at (γ_0,λ_0), where γ_0 is the (complex) solution to $\frac{d\lambda}{d\gamma} = 0$, i. e.

$$(\gamma_0 - 1)^2 + [2\gamma_0 - 1 + a\gamma_0(\gamma_0 - 1)]e^{-a\gamma_0} = 0 \qquad \text{(IV.2.8)}$$

and $\lambda_0 \equiv \lambda(\gamma_0)$. In Table IV.1 the solution $\gamma_0 = |\gamma_0|e^{i\theta_0}$ to eq. (IV.2.8) is reported for several values of the parameter a, along with the associated value of $\lambda_0 = |\lambda_0|e^{i\varphi_0}$.

Table IV.1 - Location of the branch point (γ_0,λ_0) of $\gamma(\lambda)$, eq. (IV.3.2).

a	$\|\gamma_0\|$	θ_0	$\|\lambda_0\|$	φ_0
0.1	37.03991	135.8241	32.06103	117.4178
0.5	4.41124	111.5265	5.08384	84.0719
1.0	1.68953	81.9146	2.53663	62.9295
2.0	1.02378	35.1635	1.46939	38.2273
4.0	0.99425	8.8326	1.08047	14.8016
8.0	0.99984	1.0566	1.00269	2.0904
∞	1.0	0.0	1.0	0.0

It should be noted how the convergence radius is always greater than unity, which is here the largest value of physical interest.

We shall now consider the important computational investigation by Chałasinski et al. [64] on the convergence properties of the RSPT series when applied to the case of the interaction $H\ldots H^+$ between a (ground state) hydrogen atom and a proton.

Let the Hamiltonian operator $\hat{H}$ of the "complex" H_2^+ be partitioned as $\hat{H} = \hat{H}_0 + \hat{V}$, where $\hat{H}_0 = -\frac{1}{2}\nabla^2 - \frac{1}{r_a}$ is the hydrogen atom Hamiltonian, with ground state wavefunction $|1s_a\rangle$ and energy $E_0 = -\frac{1}{2}$ a.u., and $\hat{V} = \frac{1}{R} - \frac{1}{r_b}$ is the perturbation operator corresponding to proton at a distance R from the atom. Clearly $[\hat{I},\hat{H}] = 0$, $\hat{I}$ being the operator which inverts the electron coordinates through the midpoint of the internuclear axis, but $[\hat{I},\hat{H}_0] \neq 0$. The gerade (g) or ungerade (u) symmetry of the eigenstates to $\hat{H}$ must therefore be generated by the RSPT procedure, starting from eigenstates to $\hat{H}_0$ which do not possess such symmetry.

Table IV.2 - Convergence of the Rayleigh-Schrödinger perturbation expansion for the ground state of H_2^+

n	R = 3.0 a.u. E_{0n}	R = 3.0 a.u. $\varepsilon(n)^a$	R = 12.5 a.u. E_{0n}	R = 12.5 a.u. $\varepsilon(n)^a$
1	3.3050 (-3)	104.26	1.500 (-11)	100.000
2	-2.5686 (-2)	71.14	-9.4250 (-5)	27.809
3	-1.1074 (-2)	56.87	-1.2599 (-6)	26.844
4	-9.8501 (-3)	44.17	-1.5711 (-7)	26.723
5	-8.0099 (-3)	33.84	-1.5409 (-8)	26.711
6	-6.7955 (-3)	25.08	-3.8934 (-9)	26.708
7	-5.5897 (-3)	17.87	-1.9520 (-9)	26.707
8	-4.5279 (-3)	12.03	-1.5287 (-9)	26.706
9	-3.5621 (-3)	7.45	-1.4021 (-9)	26.705
10	-2.7177 (-3)	3.94	-1.3565 (-9)	26.704
12	-1.3808 (-3)	-0.41	-1.3304 (-9)	26.702
14	-4.9101 (-4)	-2.18	-1.3254 (-9)	26.700
16	2.3496 (-5)	-2.40	-1.3244 (-9)	26.698
18	2.5690 (-4)	-1.85	-1.3241 (-9)	26.696
20	3.0523 (-4)	-1.07	-1.3241 (-9)	26.694
25	1.1249 (-4)	0.22	-1.3241 (-9)	26.688
30	-3.6059 (-5)	0.30	-1.3241 (-9)	26.683
35	-3.7218 (-5)	0.02	-1.3241 (-9)	26.678
38	-1.3906 (-5)	-0.07	-1.3241 (-9)	26.675

[a]The values of $\varepsilon(n)$ show the percentage error committed when the polarization expansion is truncated after the n-th order: $\varepsilon(n) = \frac{100}{\Delta E^g}(\Delta E^g - \sum_{k=1}^{n} E_{0k})$

Extremely accurate calculations carried out through the 38-th order of perturbation (!) have led to the results collected in Table IV. 2, where the convergence behaviour of the interaction energy for the ground state can be followed in a detailed way. The two internuclear distances considered (R = 3 a.u. and 12.5 a.u.) represent typical separations in the short and intermediate regions, respectively: in particular, R = 12.5 a.u. corresponds approximately to the internuclear separation where the excited state $|2p\sigma_u\rangle$ exhibits a shallow minimum [65], with interaction energy $\Delta E^u = -6.0774 \times 10^{-5}$ a.u.

At R = 3.0 a.u. the polarization approximation is seen to be convergent to the energy of the ground state $|1s\sigma_g\rangle$ of H_2^+, even though the convergence rate is too slow for practical applications. At R = 12.5 a.u. the situation is still worse, in the sense that the convergence rate of the series initially is fast, but drastically deteriorates, so that almost inappreciable progresses are apparent from n = 3 ÷ 4 on. As a comment, we immediately realize that i) the present behaviour agrees with our finding about the convergence radius of the RSPT series in the delta-function model of one-electron diatoms; and ii) at large internuclear distances, a RSPT procedure truncated after a few orders at first sight could appear to have practically converged. That reached is, however, only an apparent limit, because the true value is approached with exasperating slowness. The apparent limit is the <u>Coulomb energy</u> [2,6,66], i.e. $Q = (\Delta E^g + \Delta E^u)/2$, the arithmetic average of the interaction energies of the states $|1s\sigma_g\rangle$ and $|2p\sigma_u\rangle$ at R = 12.5 a.u., which corresponds to 73.275% of the interaction energy of the gerade state, while the <u>exchange energy</u> [2,6,66] $K = (\Delta E^g - \Delta E^u)/2$ is practically missing (see also next section).

The reason for the RSPT to converge so slowly is easy to understand. At large R values, the exact ground state eigenstate to $\hat{H}$ approaches $(2)^{-\frac{1}{2}}[|1s_a\rangle + |1s_b\rangle]$ and must be generated by the expansion in eigenstates $|n_a \ell_a m_a\rangle$ to $\hat{H}_0$, i.e. $|1s_b\rangle \equiv |1_b 0_b 0_b\rangle = \sum_{n_a \ell_a m_a} |n_a \ell_a m_a\rangle\langle n_a \ell_a m_a | 1s_b\rangle$, that surely requires the sum of very many tiny contributions.

IV.3. "Polarization Approximation" and its Inadequacy for Evaluating Intermolecular Interactions: Claverie's Analysis.

If we are disposed to generalize to many-electron cases the results found in the previous section for some very simple systems susceptible of a convincing investigation unobscurated by approximations of any kind, we should conclude that discarding the standard RSPT for intermolecular interaction calculations is dictated by practical requirements more than first principles. However, if we try to answer the first question raised at the end of section IV.1, we shall bring to light deeper reasons which make the "polarization approximation" procedure generally inadequate for the purposes at our heart.

Instead of becoming involved into a general analysis, we shall examine the case of a system composed by two interacting 2-electron subsystems (like He...He, H_2...H_2), because in spite of its simplicity it exhibits sufficient elements of complexity proper of more complicate situations. The present analysis will make explicit use of the results obtained in sec. II.5, leaning heavily on the example from the S_4 group there given.

We start by observing that the full Hamiltonian operator $\hat{H}(1,2,3,4)$ of the system is invariant under the symmetric group S_4, so that its space eigenstates must transform under permutations as some irreducible representation of S_4 itself, i.e. according to the Young diagrams [4], [3,1], [2^2], [$2,1^2$], [1^4], of dimensions 1,3,2,3,1 respectively (see sec. II.3). The allowed spin functions however belong to the Young diagrams [4], [3,1] and [2^2], so that the physically admissible levels correspond to space functions which transform only according to the (associate) graphs [1^4], [$2,1^2$], [2^2] (see sec. II.4). These eigenstates to $\hat{H}$ are to be produced, by standard RSPT, in the Hilbert space spanned by eigenstates to the unperturbed Hamiltonian operator $\hat{H}_0(1,2,3,4) = \hat{H}_A(1,2) + \hat{H}_B(3,4)$ which, as already remarked in sec. IV.1, is invariant under the permutations of the subgroup $S_2 \otimes \overline{S}_2$ of S_4. Since the space eigenstates to both $\hat{H}_A(1,2)$ and $\hat{H}_B(3,4)$ transform under the permutations of S_2 according to the one-dimensional representations associated with the Young graphs [2] (symmetric eigenstates) and [1^2] (antisymmetric eigenstates), the space eigenstates to $\hat{H}_0(1,2,3,4)$, which are products of space eigenstates to $\hat{H}_A(1,2)$ and $\hat{H}_B(3,4)$, may be symmetry-adapted to one of the four one-dimensional representations of $S_2 \otimes \bar{S}_2$, i.e. $[2] \otimes [2]$, $[2] \otimes [1^2]$, $[1^2] \otimes [2]$, $[1^2] \otimes [1^2]$.

As the corresponding subspaces are non-interacting with respect to $\hat{H}(1,2,3,4)$, we may work separately in each of them and in a given subspace, corresponding to a representation γ^α of $S_2 \otimes \bar{S}_2$, the "polarization approximation" procedure will generate eigenstates to $\hat{H}(1,2,3,4)$ belonging to those representations Γ^β of S_4 which are irreducible components of the induced representation $\gamma^\alpha \uparrow S_4$. As an example, in the subspace of product eigenstates belonging to $[2] \otimes [2]$, only eigenstates to the full Hamiltonian $\hat{H}(1,2,3,4)$ belonging to the S_4 representations $[4]$, $[3,1]$ and $[2^2]$ will be produced (see Table II.2), while in the subspace of functions belonging to $[1^2] \otimes [1^2]$ only eigenstates to $\hat{H}$ corresponding to the symmetries $[2^2]$, $[2,1^2]$ and $[1^4]$ are obtainable.

If the RSPT procedure starts from the ground state of $\hat{H}_0(1,2,3,4)$ (i.e. $\lambda = 0$), whose space eigenstate transforms according to $[2] \otimes [2]$, only "output" eigenstates to $\hat{H}$ ($\lambda = 1$), which transform according to $[4]$, $[3,1]$ and $[2^2]$, can be reached, so that it becomes essential to establish which final state (eigenstate to $\hat{H} = \hat{H}_0 + \hat{V}$) is (continuously) connected with the initial one (eigenstate to $\hat{H}_0$) because only such a state is within the range of the RSPT expansion (if convergent) around the point $\lambda = 0$.

In order to proceed further, we need an important theorem about the number of nodes characterizing the eigenfunctions to any Hamiltonian operator $\hat{H}$. If we arrange the eigenvalues to $\hat{H}$ in order of increasing magnitude, it can be shown that the nodes of the n-th (space) eigenfunction to $\hat{H}$ divide the domain of the variables into no more than n subdomains [67], so that the <u>ground</u> state (space) eigenfunction can have <u>no</u> nodes and therefore <u>cannot</u> be degenerate [68,69]. As a consequence, the ground state of a N-electron system must be associated with a nodeless, i.e. totally symmetric function, which is however inadmissible on physical grounds for $N>2$, as already stated (Young diagrams having more than two columns). This "true" ground state therefore is a <u>mathematical ground state</u>, to be clearly distinguished from the <u>physical ground state</u>, which is instead the lowest of the physically admissible states.

Coming back to our actual case, we see that the mathematical ground state to $\hat{H}(1,2,3,4)$ has the permutational symmetry corresponding to the Young graph $[4]$, and we may wonder whether the physical (also mathematical) ground state of the non-interacting subsystems, corresponding to $\lambda = 0$, is connected at $\lambda = 1$ (i.e. fully interacting subsystems) with the mathematical ground state of the complex or with a different state, necessarily of higher energy.

In Figs. IV.1 (a), (b) we have qualitatively represented the behaviour of the energy E, against the "coupling" parameter λ, in the two events above pointed out.

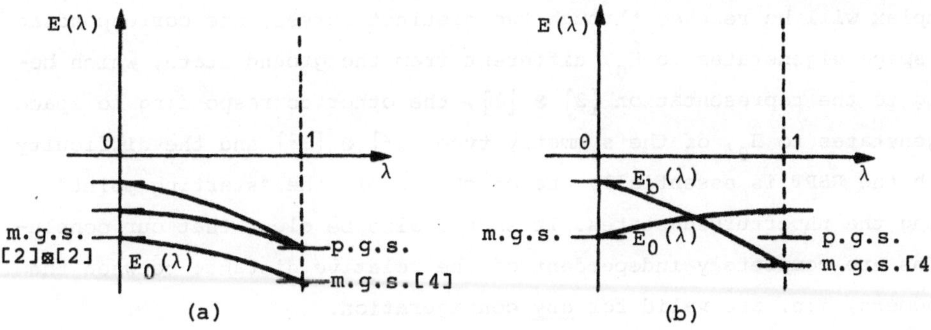

<u>Figure IV.1</u>. - Possible connections between the levels of $\hat{H}_0$ and $\hat{H} = \hat{H}_0 + \hat{V}$ (m.g.s. and p.g.s. are abridgments for mathematical ground state and physical ground state, respectively).

The second possibility, corresponding to the behaviour sketched in (b), is to be rejected. In such a case, in fact, there should exist, besides $E_0(\lambda)$, another energy function $E_b(\lambda)$ characterized by $E_b(1) \equiv$ energy of the mathematical ground state of the complex and $E_b(0) \equiv$ energy of some state of $\hat{H}_0$, other than the ground state (because this state is not degenerate and corresponds to $E_0(0)$), so that $E_0(0) < E_b(0)$, $E_0(1)$

$>E_b(1)$, that involves at least one crossing point of the two curves. Now, according to the non-crossing theorem [70], two energy curves depending on a single parameter and corresponding to levels which belong to the same symmetry cannot cross each other. Since i) only one parameter (λ) is involved, ii) the curve $E_0(\lambda)$ corresponds to the representation $[2] \otimes [2]$ (it passes through the mathematical ground state of $\hat{H}_0$) and iii) $E_b(\lambda)$ also corresponds to $[2] \otimes [2]$, because the representation $[4]$ can only be realized starting from $[2] \otimes [2]$, (see Table II.2), the conditions of validity of the non-crossing theorem are verified, which proves the impossibility of the occurrence (b).

As a conclusion, in the case of two-interacting 2-electron systems, the "polarization approximation" can only lead to the mathematical ground state of the complex (Young diagram $[4]$) instead of the physical ground state, very presumably associated with the Young diagram $[2^2]$. If we accept this very likely assumption, the physical ground state of the complex will be reached through two distinct curves, one corresponding to space eigenstates to $\hat{H}_0$, different from the ground state, which belong to the representation $[2] \otimes [2]$, the other corresponding to space eigenstates to $\hat{H}_0$, of the symmetry type $[1^2] \otimes [1^2]$ and the difficulty with the RSPT is essentially one of choice of the "starting point" among the unperturbed states. It should also be clear that our conclusions are completely independent of the relative distance between the monomers, i.e. are valid for any configuration.

If we now consider the case of two interacting atomic systems and the change of the energy of the mathematical ground state and the physical ground state of $\hat{H}$ as the internuclear distance R is varied, we get a behaviour of the kind shown in Fig. IV.2. The curve relative to the energy of the mathematical ground state is, at any R value, the lowest one and tends, as $R \to \infty$, to the sum of the ground state energies of the two separate subsystems, the same limit approached by the curve corresponding to the energy of the physical ground state. Such a behaviour is reminiscent of that displayed by the lowest singlet and triplet states of the H_2 molecule. Here the singlet plays the same role as the mathematical ground state, but it is now admissible, because only two electrons are involved; the triplet corresponds to the physi-

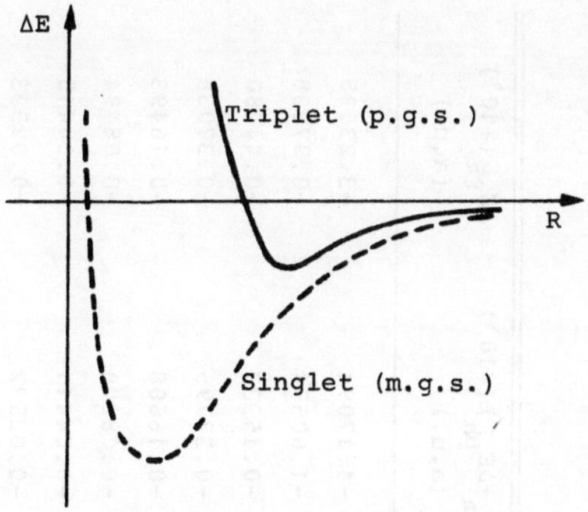

Figure IV.2. - Qualitative behaviour of the lowest singlet and triplet energy curves of two interacting hydrogen atoms as the interatomic distance R is varied. The abridgements (m.g.s.) and (p.g.s.) emphasize the analogy with the results for more complicate systems (see text).

cal ground state discussed above, in the sense that the related space eigenfunction is not totally symmetric. If this analogy is pushed further, an exponential decrease of the difference between mathematical and physical ground state energies can be expected as R increases: in fact, for two hydrogen atoms, the asymptotic behaviour of the difference $\Delta E(\text{triplet}) - \Delta E(\text{singlet}) \equiv \Delta E(\text{p.g.s.}) - \Delta E(\text{m.g.s.})$ is expressed by [6,71]

$$\Delta E(\text{triplet}) - \Delta E(\text{singlet}) \propto R^{5/2}\, e^{-2R} + O(R^2\, e^{-2R})$$

This important suggestion by Claverie actually is not a simple conjecture: Ahlrichs has been able to show that the "exchange terms", i.e. the difference between mathematical and physical ground state energies, vanish faster than any power of 1/R [72], so that the long-range portion of the physical curve, which drops more slowly (like R^{-m}) to its limit value, is not different, in a distinguishable measure, from the

Table IV.3 - Accurate energy quantities for the (mathematical) $X^1\Sigma_g^+$ and physical $b^3\Sigma_u^+$ states of H_2 at large distances (Born-Oppenheimer approximation).

R (a.u.)	$\Delta E_m (\times 10^4)$ (a.u.)	$\Delta E_{ph} (\times 10^4)$ (a.u.)	$\frac{1}{2}(\Delta E_m - \Delta E_{ph})(\times 10^4)$ (a.u.)	$\frac{1}{2}(\Delta E_m + \Delta E_{ph})(\times 10^4)$ (a.u.)	$\Delta E_{\ell r} (\times 10^4)$ (a.u.)
6.	-8.34208	+0.00019	-4.17113	-4.17094	-3.23535
7.	-1.97372	-0.03776	-0.96798	-1.00574	-0.97238
7.8	-0.70204	-0.20462	-0.24871	-0.45333	-0.44280
8.	-0.55444	-0.20147	-0.17648	-0.37795	-0.37038
9.	-0.19725	-0.13491	-0.03117	-0.16608	-0.16493
10.	-0.08738	-0.07663	-0.00537	-0.08200	-0.08194
11.	-0.04502	-0.04320	-0.00091	-0.04411	-0.04415
12.	-0.02548	-0.02516	-0.00016	-0.02532	-0.02533

mathematical one. We are therefore led to expect a practical utility of the RSPT, even though its starting point (unperturbed ground state eigensolution to $\hat{H}_0$) is not correctly connected with the desired physical ground state of the complex. However this utility is restricted to the portion of the energy curve corresponding, in general way, to very weakly interacting subsystems.

In Table IV.3 some energy quantities concerning the H_2 molecule have been collected, at different internuclear separations R [66]. ΔE_m and ΔE_{ph} represent the interaction energy of the molecule in the $X^1\Sigma_g^+$ and $b^3\Sigma_u^+$ states, respectively (in the Born-Oppenheimer approximation). The distance R = 7.8 a.u. has been included in the list because the energy curve for the excited state $b^3\Sigma_u^+$ shows a minimum around such R value. The subscripts "m" and "ph" which label ΔE_m and ΔE_{ph} are abridgments for "mathematical" and "physical" respectively, in line with our preceding remarks. The quantities

$$Q = \frac{1}{2}(\Delta E_m + \Delta E_{ph})$$
$$K = \frac{1}{2}(\Delta E_m - \Delta E_{ph}) \tag{IV.3.1}$$

included in the Table will be recognized respectively as the Coulomb and exchange energies, [2,6,66], introduced in the previous sections. It can be observed, in accordance with our expectation, that the exchange energy K decreases rather rapidly with the distance R, so that, at fairly large intermolecular separations, the difference between interaction energies associated with mathematical ground state and physical ground state curves becomes negligible. The entries $\Delta E_{\ell r}$ reported in the last column are long-range values for the interaction (dispersion) energy, obtained by approximating the interaction Hamiltonian with its multipole expansion (see next chapters). For the present purposes, we can limit ourselves to state that here $\Delta E_{\ell r}(R) = -\sum_{n=3}^{6} C_{2n} R^{2n}$, C_{2n} being van der Waals (dispersion) coefficients. As clearly seen, the more R increases, the more $\Delta E_{\ell r}$ approaches closely $Q = \frac{1}{2}(\Delta E_m + \Delta E_{ph})$, the Coulomb portion of the interaction energy. At very large distances however, $\Delta E_m \sim \Delta E_{ph} \sim Q$, so that $\Delta E_{\ell r}$ provides there an effective representation of the interaction energy.

V. SYMMETRY-ADAPTED PERTURBATION THEORIES AT LOW ORDERS: FROM H_2^+ TO THE GENERAL CASE.

V.1. The Interaction $H...H^+$ as a Test for Symmetry-Adapted Perturbation Theories.

In the previous chapter we have been able to conclude that a perturbation procedure founded on the "polarization approximation" is an inadequate tool for the evaluation of intermolecular interaction energies, its utility being possibly confined to "monomer" separations corresponding to very weakly interacting subsystems (long-range interactions).

The symmetry-adapted perturbation schemes introduced in Chap. III (and in Appendix A) are mathematical structures where the relevant symmetries of the eigenstates to the full Hamiltonian operator $\hat{H} = \hat{H}_0 + \hat{V}$ are <u>explicitly forced</u> into, from the very outset, so that the desired final states are warranted; in particular, in the case of many-electron systems, where the paramount role is played by the permutational symmetry, the proper symmetry forcing will permit to discard Pauli not-allowed eigenstates.

These remarks prescind from possible convergence problems of the chosen (symmetry-adapted) perturbation series. It is obvious, however, that <u>practical</u> reasons based on convergence rate as well as the relative difficulty of evaluation of the retained contributions will be decisive elements in dictating the choice of this or that formalism. In view of the troubles involved in the accurate solution of any actual problem, the amount of information collected till to-day is not very abundant, being really exhaustive anly for the interactions $H...H^+$ and (in minordegree) H...H.

Although an "extrapolation" founded uniquely on the simple case $H...H^+$ look a rather bold procedure, our program does rely on such assumption: the examination of a number of results relative to more complicate interacting systems (our final task in these notes) suggests *a posteriori* the reasonableness of our confidence.

A lot of attention has been devoted to the perturbative study of the interaction $H...H^+$ [40,64,73-83]. If we assume a ground state hydrogen atom (nucleus centered at a) as unperturbed system, while a proton at b is the perturbing agent (see Fig. V.1), then the unperturbed Schrödinger equation is

$$[\hat{H}_0 - E_0]|1s_a\rangle = 0$$

where

$$\hat{H}_0 = -\frac{1}{2}\nabla^2 - \frac{1}{r_a} \quad , \quad E_0 = -\frac{1}{2} \quad , \quad 1s_a \equiv \langle r_a,\theta_a,\varphi_a|1s\rangle = \pi^{-\frac{1}{2}}e^{-r_a}$$

The perturbation $\hat{V}$ due to the proton is then

$$\hat{V} = -\frac{1}{r_b} + \frac{1}{R}$$

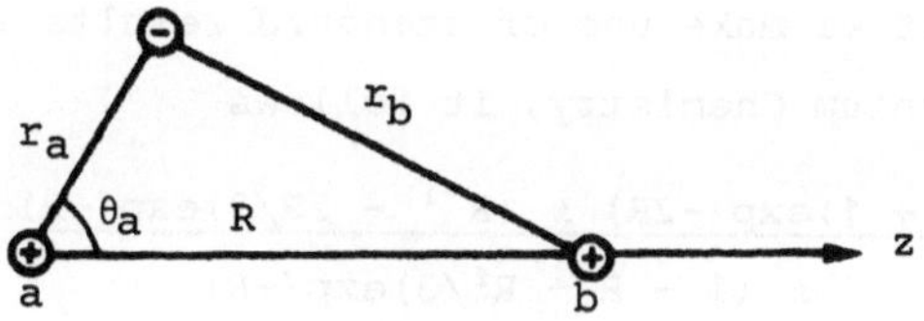

Figure V.1. - Schematic representation of the interaction $H...H^+$

As already remarked in sec. IV.2, the Hamiltonian $\hat{H} = \hat{H}_0 + \hat{V}$ commutes with the operator $\hat{I}$ that inverts the electron coordinates through the midpoint of the internuclear axis (z), so that the eigenstates to $\hat{H}$ have either gerade (g) or ungerade (u) symmetry, that must be generated starting with eigenstates to $\hat{H}_0$ which do not have such symmetry, since $[\hat{I},\hat{H}_0] \neq 0$. If we extract from the full symmetry group of $\hat{H}$ its subgroup $C_i \equiv (\hat{1},\hat{I})$, obviously isomorphic to S_2, then [eq. (II.2.5)] the proper projection operators result $\hat{Q} \equiv \hat{\rho}^{(g)} = (2)^{-1}(\hat{1}+\hat{I})$, $\hat{\rho}^{(u)} = (2)^{-1}(\hat{1}-\hat{I})$, which can be rewritten in the compact form

$$\hat{\rho}^{(i)} = \frac{1}{2}(\hat{1} + \varepsilon_i\hat{I}), \quad i \equiv g,u \begin{cases} \varepsilon_i = 1 \ , \ i \equiv g \\ \varepsilon_i = -1 \ , \ i \equiv u \end{cases} \qquad \text{(V.1.1)}$$

The 1-st order interaction energy (1-st order HL energy) is the same in all the symmetry-adapted perturbation formalisms we have considered (see eq. (III.4.1)):

$$E_{01}^{(i)} = \frac{\langle 1s_a|\hat{V}\hat{\rho}^{(i)}|1s_a\rangle}{\langle 1s_a|\hat{\rho}^{(i)}|1s_a\rangle}$$

$$= \frac{[\langle 1s_a|\hat{V}|1s_a\rangle + \varepsilon_i\langle 1s_a|\hat{V}\hat{I}|1s_a\rangle]}{1 + \varepsilon_i\langle 1s_a|\hat{I}|1s_a\rangle} \qquad \text{(V.1.2)}$$

Since $\hat{I}|1s_a\rangle = |1s_b\rangle$, the matrix element $\langle 1s_a|\hat{I}|1s_a\rangle$ at denominator is simply the overlap between two $|1s\rangle$ atomic states at the distance R; $\langle 1s_a|\hat{V}|1s_a\rangle = \frac{1}{R} - \langle 1s_a|r_b^{-1}|1s_a\rangle$ is the (algebraic) sum of the nuclear repulsion energy between the two protons and the Coulomb attractive potential energy of the (unperturbed) electronic "cloud" in the proton field. The matrix element $\langle 1s_a|\hat{V}\hat{I}|1s_a\rangle \equiv \langle 1s_a|\hat{V}|1s_b\rangle$ is the exchange integral [84,85]. If we make use of standard results to be found in any textbook on Quantum Chemistry, it follows

$$E_{01}^{(i)}(R) = \frac{(R^{-1} + 1)\exp(-2R) \pm (R^{-1} - 2R/3)\exp(-R)}{1 \pm (1 + R + R^2/3)\exp(-R)} \qquad \text{(V.1.2')}$$

the + (-) sign being associated with $i \equiv g(u)$.

Second- and higher-order energies have different forms according to the formalism employed. Their exact evaluation is a very hard task, because we are faced with the difficulty of handling the Rayleigh-Schrödinger resolvent $\hat{R}_0$, eq. (III.4.7). The standard procedure which exploits the spectral resolution of $\hat{R}_0$ (eq.(III.6.6))

$$\hat{R}_0 = \sum_{n\neq 0} (E_0 - E_n)^{-1}|\varphi_n\rangle\langle\varphi_n|$$

involves the use of the complete spectrum of eigenstates $|\varphi_n\rangle$ of the hydrogen atom, including the continuum, a true worry, so alternative approaches have long been searched for [18].

Let us consider the ket $|\varphi_{01}^{(Q)}\rangle = \hat{R}_0\hat{Q}(\hat{V} - E_{01}^{(Q)})|\varphi_0\rangle$ [see eq. (III.4.14)], which determines the 2-nd order energies in both MS-MA and EL-HAV formalism (a glance to eq. (III.4.12) is sufficient to convince that

$|\varphi_{01}^{(Q)}>$ individuates also $E_{03}^{(Q)}$ in the EL-HAV scheme). Taking into account that $(E_0 - \hat{H}_0)\hat{R}_0 = \hat{1} - |\varphi_0><\varphi_0|$, it is immediate to write down the inhomogeneous "differential equation" which admits $|\varphi_{01}^{(Q)}>$ as a particular solution:

$$(\hat{H}_0 - E_0)|\varphi_{01}^{(Q)}> = [\hat{1} - |\varphi_0><\varphi_0|]\hat{Q}(E_{01}^{(Q)} - \hat{V})|\varphi_0> \qquad (V.1.3)$$

If we are so able to find a solution to eq. (V.1.3), our problem, at the 2-nd order at least, is solved.

The 2-nd order energy in the MS-MA formalism can also be obtained starting from $|\varphi_{01}> = \hat{R}_0\hat{V}|\varphi_0>$ [see eq. (III.4.15)], a solution to the "differential equation"

$$(\hat{H}_0 - E_0)|\varphi_{01}> = [<\varphi_0|\hat{V}|\varphi_0> - \hat{V}]|\varphi_0> \equiv [E_{01} - \hat{V}]|\varphi_0> \qquad (V.1.4)$$

which appears to have a definitely simpler structure than eq. (V.1.3).

If we now specialize to the case H_2^+ tha general result expressed by eq. (V.1.3), we find

$$(\hat{H}_0 - E_0)|\varphi_{01}^{(i)}> = \frac{1}{2}[(E_{01} - \hat{V}) + \varepsilon_i(<\varphi_0|\hat{I}\hat{V}|\varphi_0> - \hat{I}\hat{V}) - \varepsilon_i E_{01}^{(i)}(<\varphi_0|\hat{I}|\varphi_0> - \hat{I})]|\varphi_0> \quad i\equiv g,u \qquad (V.1.5)$$

The solution of this equation can be reduced to that of the following set of simpler equations [37,39]

$$\begin{aligned}(\hat{H}_0 - E_0)|\varphi_{01}> &= (E_{01} - \hat{V})|\varphi_0> \\ (\hat{H}_0 - E_0)|\theta> &= (<\varphi_0|\hat{I}\hat{V}|\varphi_0> - \hat{I}\hat{V})|\varphi_0> \\ (\hat{H}_0 - E_0)|\omega> &= (\hat{I} - <\varphi_0|\hat{I}|\varphi_0>)|\varphi_0>\end{aligned} \qquad (V.1.6)$$

from which one recovers $|\varphi_{01}^{(i)}>$ as $|\varphi_{01}^{(i)}> = \frac{1}{2}[|\varphi_{01}> + \varepsilon_i|\theta> + \varepsilon_i E_{01}^{(i)}|\omega>]$.

An exact (very complicate) solution to the first of eqs. (V.1.6) is available for H_2^+ in closed form [86,87], while $|\Theta>$, in spite of noticeable efforts [75,88,89] has not been obtained in that way; the same is true for $|\omega>$. Arbitrarily accurate solutions to the three eqs. (V.1.6) are however within our reach, by finding the stationary points of the following Hylleraas-type functionals [37,39]

$$J_1[\tilde{\varphi}_{01}] = \langle\tilde{\varphi}_{01}|\hat{H}_0 - E_0|\tilde{\varphi}_{01}\rangle + 2\langle\tilde{\varphi}_{01}|\hat{V}|\varphi_0\rangle \ , \ \langle\tilde{\varphi}_{01}|\varphi_0\rangle = 0$$

$$J_2[\tilde{\theta}] = \langle\tilde{\theta}|\hat{H}_0 - E_0|\tilde{\theta}\rangle + 2\langle\tilde{\theta}|\hat{I}\hat{V}|\varphi_0\rangle \ , \ \langle\tilde{\theta}|\varphi_0\rangle = 0$$

$$J_3[\tilde{\omega}] = \langle\tilde{\omega}|\hat{H}_0 - E_0|\tilde{\omega}\rangle + 2\langle\tilde{\omega}|\hat{I}|\varphi_0\rangle \ , \ \langle\tilde{\omega}|\varphi_0\rangle = 0$$

These functionals are easily verified to have the property that, if E_0 corresponds to the energy of the ground state, then $J_1[\tilde{\varphi}_{01}] \geq J_1[\varphi_{01}]$, $J_2[\tilde{\theta}] \geq J_2[\theta]$, $J_3[\tilde{\omega}] \geq J_3[\omega]$, the equalities holding only if $|\tilde{\varphi}_{01}\rangle = |\varphi_{01}\rangle$, $|\tilde{\theta}\rangle = |\theta\rangle$ and $|\tilde{\omega}\rangle = |\omega\rangle$. The possibility of obtaining <u>approximate</u> solutions to our inhomogeneous differential equations, through the variation of parameters contained in trial functions of suitable functional forms, does not find us unprepared; the procedure, extensively exploited in perturbative calculations [18,90], clearly assumes particular importance in view of applications to many-electron situations, where exact solutions are out of question.

Highly accurate results for H_2^+ [79] obtained by using the variation procedure above, with trial functions in the form of linear combinations of Slater-type orbitals, are collected in Table V.1. $\varepsilon(n)$ is the sum of the perturbation energies through the n-th order, expressed in terms of the percentage error, i.e.

$$\varepsilon(n) = 100 \cdot \frac{(\text{exact value} - \text{approximate value})}{(\text{exact value})}$$

The values refer to the states $1s\sigma_g$ and $2p\sigma_u$, at several internuclear distances, as evaluated in different symmetry-adapted formalism: in particular, the label HS corresponds to the Hirschfelder-Silbey scheme (see Appendix A), an approach not explicitly considered in Chap. III and reported here for completeness reasons.

The quantities $\varepsilon(1)$, $\varepsilon(2)$, $\varepsilon(3)$ obtained in the "polarization approximation" (PA) have been listed because they allow a better appreciation of the results from symmetry-adapted perturbation theories. It has to be emphasized that $\varepsilon(2)$ from PA appraches the correct behaviour at large internuclear separations. $\varepsilon(1)$ from MS-MA, which is the same as $\varepsilon(1)$ from both EL-HAV and HS, is nothing but the 1-st order HL energy (see sec. III.4), in the Table indicated as HL.

The inspection of the Table for the $1s\sigma_g$ state shows that $\varepsilon(3)$ from HS is best overall, the analogous quantity from MS-MA being very good everywhere except at R = 2 a.u. (corresponding to the curve minimum). $\varepsilon(3)$ from EL-HAV is also good, except at rather large R values; in this regard, it should be pointed out the very bad behaviour of the energy through the 2-nd order from EL-HAV and the very noticeable improvement caused by 3-rd contributions. On the other hand, $\varepsilon(2)$ from MS-MA is good over the whole range of distances.

The results concerning the $2p\sigma_u$ state appear to be particularly interesting, because the behaviour of the corresponding curve is reminiscent of that of typical intermolecular interactions, as the weak attractive minimum and its location at fairly large internuclear separations ($R \sim 12.5$ a.u.) suggest. Such remark is also in line with the analysis of sec. IV.3, according to which we are led to expect that the physical ground state of a many-electron "complex" corresponds, at any separation, to some "mathematical" excited state of the full Hamiltonian operator [6,17]. The inspection of Table V.1 puts now in evidence that $\varepsilon(3)$ from HS is very good overall, the same quantity from EL-HAV being of comparable quality except at rather large R values, where the behaviour is not quantitatively adequate, in spite of the very noticeable improvement with respect to the corresponding $\varepsilon(2)$. $\varepsilon(3)$ from MS-MA is bad at small R, but good in the other regions, while $\varepsilon(2)$ in the same approximation is better than $\varepsilon(3)$ at very small distances and almost comparable elsewhere.

Since the computational burden is strongly increased as the perturbative calculations are pushed from second- to third-order, a realistic perturbative approach is unlikely to go beyond the 2-nd order. But even in this case, the search for an optimal theory, i.e. one able to produce the best results overall, is not so easy: if, however, the intrinsic simplicity of the approach is regarded as an element of merit, we feel that the MS-MA formalism is a valuable choice, which should be considered with particular attention.

Table V.1 - Interaction energy for the states $1s\sigma_g$ and $2p\sigma_u$ of H_2^+, expressed in terms of the percentage error $\varepsilon = 100$ (exact value-approximate value)/exact value (the distances are in a.u.)

Method		R=0.2	R=2.	R=4.	R=6.	R=8.	R=10.	R=12.5	R=15.
					$1s\sigma_g$				
PA	$\varepsilon(1)$	-12.6	126.8	100.9	100.1	100.0	100.0	100.0	100.0
	$\varepsilon(2)$	0.3	58.3	79.7	84.0	77.3	59.7	27.9	7.8
	$\varepsilon(3)$	-0.1	35.7	71.9	81.4	76.0	58.6	26.8	7.1
HL	$\varepsilon(1)$	-12.5	47.6	20.0	24.2	32.7	48.6	76.5	93.8
MS-MA	$\varepsilon(1)$	-12.5	47.6	20.0	24.2	32.7	48.6	76.5	93.8
	$\varepsilon(2)$	0.4	-2.7	0.8	2.7	3.6	3.1	1.9	1.0
	$\varepsilon(3)$	-0.2	11.5	0.9	0.4	1.0	0.9	0.5	0.1
EL-HAV	$\varepsilon(2)$	0.5	-6.9	1.9	5.1	7.5	10.8	16.7	20.6
	$\varepsilon(3)$	0.0	2.6	0.3	1.1	1.8	2.5	3.8	4.6
HS	$\varepsilon(2)$	0.5	-5.6	1.2	3.3	3.8	3.2	1.9	1.0
	$\varepsilon(3)$	-0.1	4.2	0.7	1.1	1.2	0.9	0.5	0.1
Exact energy (a.u.)[a]		3.5714	-0.10263	-0.4608(-1)	-0.1197(-1)	-0.2570(-2)	-0.5787(-3)	-0.1306(-3)	-0.4894(-4)

Table V.1 (continued) - Interaction energy for the states $1s\sigma_g$ and $2p\sigma_u$ of H_2^+, expressed in terms of the percentage error: 100 (exact value-approximate value)/exact value (distances in a.u.)

Method		R=0.2	R=2.	R=4.	R=6.	R=8.	R=10.	R=12.5	R=15.
					$2p\sigma_u$				
HL	$\varepsilon(1)$	-13.4	-2.0	-1.2	-6.7	-26.6	-202.0	150.6	107.2
MS-MA	$\varepsilon(1)$	-13.4	-2.0	-1.2	-6.7	-26.6	-202.0	150.6	107.2
	$\varepsilon(2)$	-32.9	-6.0	1.3	1.7	1.9	2.9	0.6	0.7
	$\varepsilon(3)$	-110.4	-10.1	1.3	1.2	1.1	2.3	-0.3	0.0
EL-HAV	$\varepsilon(2)$	10.0	-0.4	-0.1	-1.1	-5.2	-42.0	32.5	23.5
	$\varepsilon(3)$	2.3	-0.1	-0.0	-0.2	-1.1	-9.1	7.2	5.3
HS	$\varepsilon(2)$	0.6	-1.9	0.7	1.3	1.8	2.9	0.6	0.7
	$\varepsilon(3)$	16.3	-0.3	0.1	0.5	0.8	2.2	-0.3	0.0
Exact energy (a.u.)[a]		4.9973	0.33247	0.5445(-1)	0.9356(-2)	0.1394(-2)	0.9893(-4)	-0.6077(-4)	-0.4206(-4)

(a) See ref. [65].

V.2 - Some Useful Concepts Arising from an Analysis of the Interaction $H...H^+$.

We now propose to emphasize a number of points, some of which ensue from an analysis of the results presented in the preceding section. As some conclusions drawn from our present "simple" system will be transferable to many-electron situations, they should be considered with attention.

We start by investigating a simple approximation to MS-MA $\varepsilon(2)$, characterized with respect to such approach by a lower amount of computational difficulties, so that it could possibly be of interest in view of applications to more complicate interactions than $H...H^+$. $\varepsilon(2)$ in this approximation consists of the sum of the 1-st order HL energy $E_{01}^{(i)}$ and the 2-nd order energy E_{02} evaluated in the "polarization approximation", the so-called 2-nd order polarization energy. The resulting quantity is reported in Table V.2 for both states $1s\sigma_g$ and $2p\sigma_u$ at several internuclear distances (note that E_{02} is the same for both states). It is immediately apparent how this approximation, although not completely satisfactory from a quantitative point of view, is eminently reasonable over the whole range of internuclear separations for $1s\sigma_g$ and $2p\sigma_u$ (the most conspicuous errors occurring at the very critical distances where the interaction energy changes sign).

We can appreciate the terms missing in $E_{01}^{(i)} + E_{02}$ with respect to MS-MA $\varepsilon(2)$ by properly rearranging the expression of $E_{02}^{(i)}$, eq. (III.4.9). If we insert $\hat{Q} \equiv \hat{\rho}^{(i)} = (2)^{-1}[1 + \varepsilon_i \hat{I}]$ and make use of eq. (V.1.2) for $E_{01}^{(i)}$, by simple manipulations we find

$$E_{02}^{(i)} = E_{02} + \varepsilon_i \frac{<\varphi_{01}|\hat{I}\hat{V}|\varphi_0> - E_{01}<\varphi_{01}|\hat{I}|\varphi_0> - E_{02}S}{1 + \varepsilon_i S} - \frac{<\varphi_0|\hat{V}\hat{I}|\varphi_0> - E_{01}S}{(1 + \varepsilon_i S)^2}<\varphi_{01}|\hat{I}|\varphi_0> \qquad (V.2.1)$$

where we have introduced the notation $S = <\varphi_0|\hat{I}|\varphi_0>$ for the overlap intergral between $|\varphi_0> \equiv |1s_a>$ and $\hat{I}|\varphi_0> \equiv |1s_b>$. Clearly $E_{01} = <\varphi_0|\hat{V}|\varphi_0>$, the 1-st order perturbation energy in the "polarization approximation", often referred to as the coulombic or electrostatic part of $E_{01}^{(i)}$, while

$|\varphi_{01}\rangle = \hat{R}_0 \hat{V} |\varphi_0\rangle$.

Table V.2 - Interaction energy for the states $1s\sigma_g$ and $2p\sigma_u$ of H_2^+ from a simplified version of the MS-MA approach. The energy quantities are expressed in terms of the percentage error ε=100·(exact value -approximate value)/exact value.

Method		R=2.	R=4.	R=6.	R=8.	R=10.	R=15.
				$1s\sigma_g$			
MS-MA	$\varepsilon(2)$	-2.7	0.8	2.7	3.6	3.1	1.0
$E_{01}^{(g)} + E_{02}$		-20.8	-1.2	8.1	10.1	8.3	1.6
				$2p\sigma_u$			
MS-MA	$\varepsilon(2)$	-6.0	1.3	1.7	1.9	2.9	0.7
$E_{01}^{(u)} + E_{02}$		19.1	16.7	14.0	15.2	33.8	-0.1

One sees therefore that $E_{01}^{(i)} + E_{02}$ differs from MS-MA $\varepsilon(2)$ by the last two terms on the right-hand side of eq. (V.2.1), which vary as different powers of the overlap S. In particular, the second term, which can be rewritten as

$$E_{02,ex}^{(i)} = \varepsilon_i \frac{\langle\varphi_{01}|(\hat{I} - S\hat{1})(\hat{V} - E_{01}\hat{1})|\varphi_0\rangle}{1 + \varepsilon_i S} \qquad \text{(V.2.2)}$$

for small S values has a magnitude order equal to the overlap itself; following the suggestion by Murrell *et al.* [36,91], $E_{02,ex}^{(i)}$ will be called <u>exchange polarization energy</u>. The denomination proposed should not be interpreted too literally, in the sense there is actually a physically observable exchange process of polarization between the two nuclear environments in $H...H^+$. What we intend to say is that the presence of the proton distorts the atomic wavefunction $|1s_a\rangle$ centered at the nucleus a and an analogous distortion must reveal itself also at the nucleus b, as a consequence of the symmetry under permutation of the two nuclei displayed by the Hamiltonian $\hat{H}$ of the system [81].

Some other interesting and important results follow if we analyze the perturbation operator $\hat{V} = R^{-1} - r_b^{-1}$ in terms of spherical harmonics centered at the nucleus a, i.e. perform a so-called multipole expansion of the potential energy operator $\hat{V}$ according to

$$\hat{V} = \sum_{\ell=0}^{\infty} V_\ell(r_a,R)\, P_\ell(\cos\theta_a) \qquad (V.2.3)$$

where $P_\ell(\cos\theta_a)$ is the ℓ-th Legendre polynomial of the indicated argument. By exploiting the orthogonality of the Legendre polynomial set, it is not difficult to show that $V_\ell(r_a,R) = (R^{-1}\delta_{\ell 0} - r_<^\ell/r_>^{\ell+1})$, $r_<(r_>)$ being the lesser (greater) of r_a and R. An entirely equivalent, but more useful form for $V_\ell(r_a,R)$ is given by [81],

$$V_\ell(r_a,R) = V_\ell^{as}(r_a,R) + V_\ell^{pen}(r_a,R) \qquad (V.2.4)$$

where

$$V_\ell^{as}(r_a,R) = \begin{cases} 0 & \ell = 0 \\ -\dfrac{r_a^\ell}{R^{\ell+1}} & \ell \neq 0 \end{cases}$$

$$V_\ell^{pen}(r_a,R) = \Theta(r_a - R)\left[\frac{r_a^\ell}{R^{\ell+1}} - \frac{R^\ell}{r_a^{\ell+1}}\right]$$

having introduced the Heaviside step-function $\Theta(x) = 1$, $x>0$, $\Theta(x) = 0$, $x \leq 0$. $V_\ell^{as}(r_a,R)$ characterizes the asymptotic behaviour of the ℓ-th term of the multipolar expansion (V.2.3) and is actually the only retained in the usual long-range treatment of intermolecular interactions (see next chapters). $V_\ell^{pen}(r_a,R)$, the commonly neglected component, takes into account penetration or charge-overlap effects, i.e. the fact that the proton is actually embedded within the electron charge distribution. If we express the solution $|\varphi_{01}\rangle$ to eq. (V.1.4) in form

$$|\varphi_{01}\rangle = \sum_{\ell=0}^{\infty} |\varphi_{01}^{(\ell)}\rangle$$

the 2-nd order polarization energy can be written as a multipolar expansion,

$$E_{02} = \sum_{\ell=0}^{\infty} E_{02}^{(\ell)} \qquad \text{(V.2.5)}$$

where

$$E_{02}^{(\ell)} = \langle \varphi_{01}^{(\ell)} | \hat{V} | \varphi_0 \rangle \qquad \text{(V.2.6)}$$

$E_{02}^{(\ell)}$ represents the interactione energy between the electric charge of the proton at b and the <u>induced</u> electric 2^{ℓ}-pole in the charge distribution of the atom centered at a, so we usually speak of 2-nd order <u>induction energy</u> (we shall see in the next section how in general the polarization energy includes also <u>dispersion energy</u>).

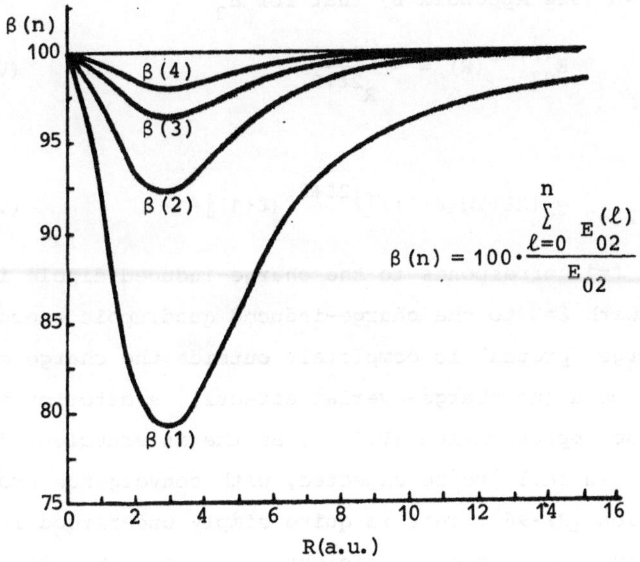

<u>Figure V.2</u>. - Comparison between 2-nd order induction energy and its approximations $E_{02}(n) = \sum_{\ell=0}^{\infty} E_{02}^{(\ell)}$ for H_2^+ as a function of the nuclear separation R.

In Fig. V.2 are reported the results obtained by Kr ek and Meath [92] for H_2^+ as a test of convergence for the series (V.2.8). The curves are a plot of the ratio $\beta(n) = 100 \cdot \sum_{\ell=0}^{n} E_{02}^{(\ell)} / E_{02}$ as a function of the internuclear separation R and put in evidence a rather smooth convergence to E_{02} over the whole range of distances.

A very interesting approximation to E_{02} (with possible practical consequences) is obtained by completely neglecting the penetration effects of the proton in the electronic charge "cloud", so that $V_\ell(r_a,R) \simeq V_\ell^{as}(r_a,R)$, $|\varphi_{01}\rangle \simeq |\varphi_{01}^{as}\rangle$ and

$$E_{02} \simeq E_{02}^{as} = \sum_{\ell=1}^{\infty} E_{02}^{(\ell)as} \qquad (V.2.7)$$

where

$$E_{02}^{(\ell)as} = \langle \varphi_{01}^{(\ell)as} | \hat{V}^{as} | \varphi_0 \rangle$$

It can be shown (see Appendix B) that for H_2^+

$$E_{02}^{(\ell)as}(R) = -\frac{C_{2\ell+2}}{R^{2\ell+2}} \qquad (V.2.8)$$

where

$$C_{2\ell+2} = (2\ell+2)!(\ell+2)/[2^{2\ell+2}\ell(\ell+1)] \qquad (V.2.9)$$

The term with $\ell=1$ corresponds to the charge-induced dipole interaction energy, that with $\ell=2$ to the charge-induced quadrupole energy, etc., when the charge (proton) is completely outside the charge distribution of the atom a (no charge-overlap effects). A deterioration of the goodness of the approximation (V.2.7), as the internuclear distance R decreases, is generally to be expected, with convergence problems worth of investigation [92-96]. This is quite simply understood if one notes the increase in size of the coefficient $C_{2\ell+2}$, eq. (V.2.9), as the power of R^{-1} increases. The series (V.2.7) is actually divergent for any R value, as easily verified by applying the ratio test. However, for any positive integer n,

$$\lim_{R\to\infty} R^{2n+2}\left[E_{02}(R) - \sum_{\ell=1}^{n} C_{2\ell+2}R^{-(2\ell+2)}\right] = 0$$

which characterizes the series (V.2.7) as an asymptotic (or semiconvergent) one [58,97]: for large R values, therefore, $E_{02}(R)$ can quite accurately be represented by such series, there being for any fixed R value an optimal number of terms that provide the best approximation.

Fig. V.3 allows to compare $E_{02}(R)$ with the results arising from some truncated asymptotic expansions $\tilde{E}_{02}(n) = -\sum_{\ell=1}^{n} C_{2\ell+2}R^{-(2\ell+2)}$: the agreement of the asymptotic series with the exact one, for large R values, is manifest, along with their bad divergent behaviour as R decreases [92].

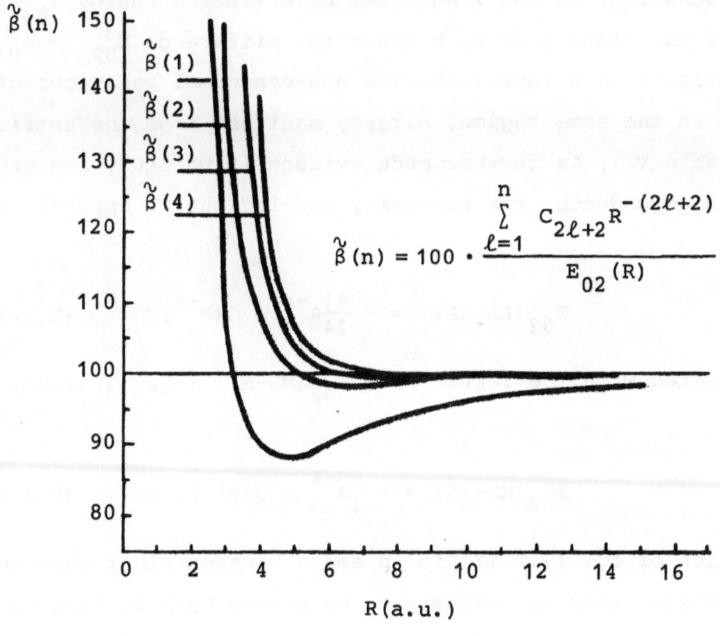

Figure V.3. - Comparison between 2-nd order induction energy and its approximation $\tilde{E}_{02}(n) = -\sum_{\ell=1}^{n} C_{2\ell+2}R^{-(2\ell+2)}$ for H_2^+ as a function of the nuclear separation R.

The same kind of multipole analysis carried out for the induction energy E_{02} could be extended rather readily to the exchange induction energy [81]; thus, in analogy to eqs. (V.2.5), (V.2.6), we have

$$E_{02,ex} = \sum_{\ell=0}^{\infty} E_{02,ex}^{(\ell)} \qquad \text{(V.2.10)}$$

where $E^{(\ell)}_{02,ex}$ may be called the <u>induced</u> <u>2^{ℓ}-pole</u> <u>exchange</u> <u>energy</u>. From an analysis of the results obtained by Chalasiński *et al.* [80,81] it is likely that in the range of distances corresponding to small overlap values one may evaluate 2-nd order exchange effects in terms of $|\varphi^{as}_{01}\rangle$ [75].

As a conclusion of the present section, we would like to discuss the bad behaviour of the long-range 2-nd order energy $E^{(i)}_{02}$ from the EL-HAV method, which appears in striking contrast with respect to the same order energy from other formalisms. As a first remark, we note from eq. (V.2.1) that $E^{(i)}_{02}$ from MS-MA is expected to have the same long-range behaviour as the 2-nd order polarization energy E_{02}, independently of the state g or u, because the difference $E^{(i)}_{02} - E_{02}$ vanish exponentially as R increases. The non-conformal behaviour of E_{02}(EL-HAV) in the same region, already manifest from the entries collected in Table V.1, is further made evident if we study its explicit long-range R dependence; the analysis, carried out in Appendix B, shows that

$$E_{02}(\text{EL-HAV}) = -\frac{41}{24}R^{-4} + O(R^{-6}) \qquad (\text{V.2.11})$$

to be contrasted with the result from E_{02}(MS-MA) [eqs. (V.2.8), (V.2.9)]

$$E_{02}(\text{MS-MA}) = -\frac{9}{4}R^{-4} + O(R^{-6}) \qquad (\text{V.2.11}')$$

The derivation of eq. (V.2.11) in Appendix B makes clear that the origin of the discrepancy is uniquely to be traced back to lack of meaning of the concept of perturbation order in symmetry-adapted perturbation theories. For this very reason, it is not absolutely amazing that the third-order contributions may influence in such noticeable way (see Table V.1) the second-order result [98,99].

It has been argued that the EL-HAV formalism can be cured for its undesired long-range behaviour at the second-order, without making recourse to the difficult evaluation of 3-rd order contributions [100]. What is needed is performing a Feenberg transformation [101,102] on the Hamiltonian operator $\hat{H}$, so that

$$\hat{H} = \hat{H}_0(\xi) + \hat{V}(\xi) \qquad \text{(V.2.12)}$$

where

$$\hat{H}_0(\xi) = \xi^{-1}\hat{H}_0$$

$$\hat{V}(\xi) = \hat{V} + (\xi - 1)\xi^{-1}\hat{H}_0$$

ξ being an arbitrary parameter. As a consequence of the invariance of the eigenstates to $\hat{H}_0$ under the transformation $\hat{H}_0 \rightarrow \hat{H}_0(\xi)$, while the corresponding eigenvalues become uniformly scaled ξ^{-1} times, it is easily verified that the sum $E_0 + E_{01}^{(Q)}$ does not change passing from the standard Rayleigh-Schrödinger partition $\hat{H} = \hat{H}_0 + \hat{V}$ to that expressed by eq. (V.2.12). Moreover, taking into account that the resolvent $\hat{R}_0$ transforms according to $\hat{R}_0 \rightarrow \xi\hat{R}_0$ and the various symmetry operators $\hat{Q}$, $\hat{P}_1$, etc. are unchanged, the energy contributions of higher order than the 1-st one become ξ-dependent; in particular, $E_{02}^{(Q)} \rightarrow \xi E_{02}^{(Q)}$. Thus, as a result of the Feenberg transformation on $\hat{H}$, we have

$$E = E_0 + E_{01}^{(Q)} + \xi E_{02}^{(Q)} + \ldots\ldots \qquad \text{(V.2.13)}$$

where $E_{02}^{(Q)}$ is the EL-HAV 2-nd order energy, eq. (III.4.11). If we are able to choose the parameter ξ in such a way that

$$\lim_{R\to\infty} \frac{\xi E_{02}^{(Q)}}{E_{02}} = 1 \qquad \text{(V.2.14)}$$

E_{02} being the 2-nd order polarization energy, our program is accomplished.

For H_2^+, a sensible value for ξ can be estimated from eqs. (V.2.11), (V.2.11'), $\xi = 54/41 \simeq 1.317$, rather distant indeed from the suggestion $\xi = N^{-1} = \langle\varphi_0|\hat{Q}|\varphi_0\rangle^{-1} \simeq 2$ to be found in the literature [41,100].

V.3 - A Basic Partition of the Interaction Energy Through the Second-Order for Many-Electron Subsystems.

The generalization to many-electron "complexes" of the same partition into Coulombic and exchange components as for $H...H^+$ is the important step we are now about to carry out. Even though the situation corresponding to an arbitrary number of interacting "monomers" is anything but academic, in the course of the present notes we shall be interested in the case of two subsystems A, B, so that our unperturbed Hamiltonian operator is $\hat{H}_0 = \hat{H}_0^A + \hat{H}_0^B$, with definite electrons allotted to A and B (see sec. IV.1). If $|\varphi_0^A\rangle$ and $|\varphi_0^B\rangle$ represent the exact ground state eigensolutions to $\hat{H}_0^A$ and $\hat{H}_0^B$, respectively, with eigenvalues E_0^A and E_0^B, then clearly

$$\begin{aligned} |\varphi_0\rangle &= |\varphi_0^A\varphi_0^B\rangle \equiv |\varphi_0^A\rangle|\varphi_0^B\rangle \\ E_0 &= E_0^A + E_0^B \end{aligned} \tag{V.3.1}$$

Before becoming involved in the true derivation, however, some remarks about the projector $\hat{Q}$ appearing in the employed symmetry-adapted perturbation theory are in order.

The relevant (and vital) symmetry to be forced into the perturbative formalism is that connected with the Pauli principle of antisymmetry, so that $\hat{Q} \equiv \hat{A} = (N!)^{-1} \sum_{j=1}^{N!} (-1)^j \hat{P}_j$, the full antisymmetrizer corresponding to the $N = N^A + N^B$ electrons of the "complex" AB (see eq. (III.3.2)). It should be noted that the permutations $\hat{P}_j$ involved in $\hat{Q}$ perform one, two, ... exchanges of spin- and space-coordinates within as well as between the two interacting partners. Besides the projector $\hat{A}$, let us consider the operator (not a projector!)

$$\hat{\mathcal{Q}}_{int} = \sum_j (-1)^{p_j} \hat{P}_j^{AB} \tag{V.3.2}$$

which sums only inter-partners permutations, weighted by the corresponding parity-depending factors $(-1)^{p_j}$. We have clearly

$$\hat{\mathcal{Q}}_{int} = \hat{1} - \hat{P}_{(1)}^{AB} + \hat{P}_{(2)}^{AB} + \dots\dots \tag{V.3.3}$$

$\hat{P}^{AB}_{(1)}$ being the sum of all permutations exchanging one pair of electron labels between the partners A, B, $\hat{P}^{AB}_{(2)}$ the sum of all permutations exchanging two pairs of electron labels between the same partners, etc. [91,103]. In explicit terms,

$$\hat{P}^{AB}_{(1)} = \sum_{i=1}^{N^A} \sum_{j=N^A+1}^{N^A+N^B} \hat{P}_{ij}$$

$$\hat{P}^{AB}_{(2)} = \sum_{i=1}^{N^A-1} \sum_{i'>i}^{N^A} \sum_{j=N^A+1}^{N^A+N^B-1} \sum_{j'>j}^{N^A-N^B} \hat{P}_{ij}\hat{P}_{i'j'} \tag{V.3.4}$$

....

If $\hat{\mathcal{Q}}^A$ ($\hat{\mathcal{Q}}^B$) is an operator formally analogous to $\hat{\mathcal{Q}}_{int}$, which exchanges only electron labels within A (B), we have

$$\hat{\mathcal{Q}}_{int}\hat{\mathcal{Q}}^A\hat{\mathcal{Q}}^B = (\sum_j (-1)^{p_j} \hat{P}^{AB}_j)(\sum_{jA} (-1)^{p_{jA}} \hat{P}_{jA})(\sum_{jB} (-1)^{p_{jB}} \hat{P}_{jB}) =$$

$$= \sum_j \sum_{jA} \sum_{jB} (-1)^{p_j p_{jA} p_{jB}} (\hat{P}^{AB}_j \hat{P}_{jA} \hat{P}_{jB})$$

But the set of all products $\hat{P}^{AB}_j \hat{P}_{jA} \hat{P}_{jB}$ is exactly the set of all permutations $\hat{P}$ of the $(N^A + N^B)$ labels $(1 \ldots N^A, N^A+1, \ldots N^A+N^B)$, so that we get [104]

$$\hat{\mathcal{Q}}_{int}\hat{\mathcal{Q}}^A\hat{\mathcal{Q}}^B = \hat{\mathcal{Q}} \tag{V.3.5}$$

where

$$\hat{\mathcal{Q}} = \sum_{j=1}^{(N^A+N^A)!} (-1)^{p_j} \hat{P}_j \tag{V.3.6}$$

In a similar way it may be shown that [104]

$$\hat{\mathcal{Q}}^A\hat{\mathcal{Q}}^B\hat{\mathcal{Q}}_{int} = \hat{\mathcal{Q}} \tag{V.3.5'}$$

Another important result that we limit ourselves to state is that both $\hat{\mathcal{Q}}^A$ and $\hat{\mathcal{Q}}^B$ commute with each of the operators $\hat{P}^{AB}_{(1)}$, $\hat{P}^{AB}_{(2)}$,... defined ie eqs. (V.3.4).

Let now $|\varphi^A_n>$ and $|\varphi^B_m>$ be any two eigenstates to $\hat{H}^A_0$ and $\hat{H}^B_0$, respectively; then,

$$\hat{A}^A|\varphi_n^A\rangle \equiv (N^A!)^{-1}\hat{\mathcal{A}}^A|\varphi_n^A\rangle = |\varphi_n^A\rangle$$

$$\hat{A}^B|\varphi_m^B\rangle \equiv (N^B!)^{-1}\hat{\mathcal{A}}^B|\varphi_m^B\rangle = |\varphi_m^B\rangle$$

because $|\varphi_n^A\rangle$ ($|\varphi_m^B\rangle$) is antisymmetric under the permutations of the antisymmetrizer $\hat{A}^A$ ($\hat{A}^B$). Therefore,

$$\hat{\mathcal{A}}_{int}|\varphi_n^A\varphi_m^B\rangle = \hat{\mathcal{A}}_{int}\hat{A}^A\hat{A}^B|\varphi_n^A\varphi_m^B\rangle = (N^A!)^{-1}(N^B!)^{-1}\hat{\mathcal{A}}_{int}\hat{\mathcal{A}}^A\hat{\mathcal{A}}^B|\varphi_n^A\varphi_m^B\rangle =$$

$$= (N^A!)^{-1}(N^B!)^{-1}\hat{\mathcal{A}}|\varphi_n^A\varphi_m^B\rangle \qquad \text{(V.3.6)}$$

having made use of eq. (V.3.5). If we divide both sides by $(N^A+N^B)!$, it follows

$$\hat{A}|\varphi_n^A\varphi_m^B\rangle = \hat{A}_{int}|\varphi_n^A\varphi_m^B\rangle \qquad \text{(V.3.7)}$$

where $\hat{A}$ is the full antisymmetrizer introduced at the beginning of the present section and

$$\hat{A}_{int} = \frac{(N^A!)(N^B!)}{(N^A+N^B)!}\sum_j (-1)^{p_j}\hat{P}_j^{AB} \qquad \text{(V.3.2')}$$

Eq. (V.3.7) establishes the important result, to be expected on intuitive grounds, that the antisymmetrization of the product $|\varphi_n^A\varphi_m^B\rangle$ of two separately antisymmetric kets requires <u>only</u> exchanges <u>between</u> the two partners $|\varphi_n^A\rangle$ and $|\varphi_m^B\rangle$.

We are now in a position to write down the Coulombic and exchange partition of the interaction energy for two subsystems A, B. If we put $\hat{P}_1^{AB} = \hat{1}$, the identity operator, and identify the projector $\hat{Q}$ with $\hat{A}_{int}$, then from eq. (V.3.2') $\hat{Q} \equiv \hat{A}_{int} = (\hat{1} + \hat{\mathcal{P}}^{AB})$, where $\hat{\mathcal{P}}^{AB} = -\hat{P}_{(1)}^{AB} + \hat{P}_{(2)}^{AB} + \ldots$ (see eq. (V.3.3)). The 1-st order HL energy $E_{01}^{(Q)}$, eq. (III.4.1), can then readily be put in the following form

$$E_{01}^{(Q)} = \langle\varphi_0|\hat{V}|\varphi_0\rangle + \frac{\langle\varphi_0|\hat{V}\hat{\mathcal{P}}^{AB}|\varphi_0\rangle - \langle\varphi_0|\hat{V}|\varphi_o\rangle\langle\varphi_o|\hat{\mathcal{P}}^{AB}|\varphi_0\rangle}{1 + \langle\varphi_o|\hat{\mathcal{P}}^{AB}|\varphi_0\rangle} =$$

$$= E_{01} + \frac{\langle\varphi_0|(\hat{V} - E_{01})(\hat{\mathcal{P}}^{AB} - \mathcal{P}_0^{AB})|\varphi_0\rangle}{1 + \mathcal{P}_0^{AB}} \equiv E_{01} + E_{01,ex} \qquad \text{(V.3.8)}$$

where $E_{01} = \langle\varphi_0|\hat{V}|\varphi_0\rangle$ is the Coulombic or electrostatic contribution to $E_{01}^{(Q)}$ (see next chapter) and $\mathcal{P}_0^{AB} = \langle\varphi_0|\hat{\mathcal{P}}^{AB}|\varphi_0\rangle$. The last term on the right-hand side of eq. (V.3.8), the exchange part of the 1-st order HL energy $E_{01}^{(Q)}$, is a basic ingredient of the interaction energy, which is out of the range of the "polarization approximation" theory truncated after a finite order. $\mathcal{P}_0^{AB}$ plays here the role of a generalized overlap; as a consequence, to the extent the two subsystems interact weakly, the importance of multiple exchange effects in $\hat{\mathcal{P}}^{AB}$ is expected to decrease more and more, with possible practical simplification of the complexity of eq. (V.3.8).

As far as the 2-nd order contribution to the interaction energy is concerned, we shall turn our attention to $E_{02}^{(Q)}$ as evaluated in MS-MA formalism, eq. (III.4.9). Also in this case, by simple manipulations we get the following generalization of eqs. (V.2.1), (V.2.2)

$$
\begin{aligned}
E_{02}^{(Q)} &= \langle\varphi_0|\hat{V}\hat{R}_0\hat{V}|\varphi_0\rangle + \frac{\langle\varphi_0|\hat{V}\hat{R}_0\hat{\mathcal{P}}^{AB}\hat{V}|\varphi_0\rangle - \mathcal{P}_0^{AB}\langle\varphi_0|\hat{V}\hat{R}_0\hat{V}|\varphi_0\rangle}{1+\mathcal{P}_0^{AB}} \\
&\quad - E_{01}\frac{\langle\varphi_0|\hat{V}\hat{R}_0\hat{\mathcal{P}}^{AB}|\varphi_0\rangle}{1+\mathcal{P}_0^{AB}} \qquad \text{(V.3.9)} \\
&= \langle\varphi_0|\hat{V}\hat{R}_0\hat{V}|\varphi_0\rangle + \frac{\langle\varphi_0|\hat{V}\hat{R}_0(\hat{\mathcal{P}}^{AB} - \mathcal{P}_0^{AB})(\hat{V} - E_{01})|\varphi_0\rangle}{1+\mathcal{P}_0^{AB}} \\
&\quad - \frac{\langle\varphi_0|\hat{V}\hat{R}_0\hat{\mathcal{P}}^{AB}|\varphi_0\rangle\langle\varphi_0|(\hat{V} - E_{01})(\hat{\mathcal{P}}^{AB} - \mathcal{P}_0^{AB})|\varphi_0\rangle}{(1+\mathcal{P}_0^{AB})^2} \\
&\simeq E_{02} + \frac{\langle\varphi_0|\hat{V}\hat{R}_0(\hat{\mathcal{P}}^{AB} - \mathcal{P}_0^{AB})(\hat{V} - E_{01})|\varphi_0\rangle}{1+\mathcal{P}_0^{AB}} = E_{02} + E_{02,ex}
\end{aligned}
$$

where $E_{02} = \langle\varphi_0|\hat{V}\hat{R}_0\hat{V}|\varphi_0\rangle$ is the 2-nd order polarization energy and

$E_{02,ex}$ the exchange polarization energy. The last expression in eq. (V.3.9), an approximation to the rigorous result of the previous line, is simply obtained by the neglect of the term with denominator $(1 + \mathcal{S}_0^{AB})^2$. Such neglected term appears to be of an order of magnitude smaller than the retained one if only single exchanges are taken into account in $\hat{\mathcal{S}}^{AB}$, so that eq. (V.3.9) in its approximate form should be very adequate in the region of small overlap values, where the two "monomers" interact weakly [105].

The 2-nd order energy expression, eq. (V.3.9), is susceptible of a further partition, of noticeable interest from both conceptual and computational standpoints. The basis of such development is offered by the following decomposition of the reduced resolvent $\hat{R}_0$,

$$\begin{aligned}\hat{R}_0 &= [(E_0^A - \hat{H}_0^A) + (E_0^B - \hat{H}_0^B)]^{-1}[\hat{1} - |\varphi_0^A\varphi_0^B\rangle\langle\varphi_0^A\varphi_0^B|] \\ &= \left(\sum_n \sum_m\right)' \frac{|\varphi_n^A\varphi_m^B\rangle\langle\varphi_n^A\varphi_m^B|}{(E_0^A - E_n^A) + (E_0^B - E_m^B)} \\ &= \hat{O}_0^A\hat{R}_0^B + \hat{O}_0^B\hat{R}_0^A + \hat{R}_0^{AB}\end{aligned} \tag{V.3.10}$$

where

$$\hat{O}_0^C = |\varphi_0^C\rangle\langle\varphi_0^C|$$

$$\hat{R}_0^C = [E_0^C - \hat{H}_0^C]^{-1}[\hat{1} - |\varphi_0^C\rangle\langle\varphi_0^C|] = \sum_{k\neq 0} \frac{|\varphi_k^C\rangle\langle\varphi_k^C|}{E_0^C - E_k^C} \tag{V.3.11}$$

$$\begin{aligned}\hat{R}_0^{AB} &= [(E_0^A - \hat{H}_0^A) + (E_0^B - E_0^B)]^{-1}[\hat{1} - \hat{O}_0^A][\hat{1} - \hat{O}_0^B] \\ &= \sum_{n\neq 0}\sum_{m\neq 0} \frac{|\varphi_n^A\varphi_m^B\rangle\langle\varphi_n^A\varphi_m^B|}{(E_0^A - E_n^A) + (E_0^B - E_m^B)}\end{aligned}$$

Evidently $\hat{O}_0^C$ is the projector on the reference space spanned by the ground state ket $|\varphi_0^C\rangle$ and $\hat{R}_0^C$ the reduced resolvent associated with the unperturbed Hamiltonian $\hat{H}_0^C$, while $\hat{R}_0^{AB}$ is the $\hat{R}_0$ component which involves only simultaneous excited eigenstates to $\hat{H}_0^A$ and $\hat{H}_0^B$.

It is now convenient to partition the interaction potential energy operator $\hat{V}$ in the form

$$\hat{V} = \hat{V}^{A}_{en} + \hat{V}^{B}_{en} + \hat{V}^{AB}_{ee} + \hat{V}^{AB}_{nn} \tag{V.3.12}$$

where

$$\hat{V}^{A}_{en} = - \sum_{i\in A} \sum_{\nu\in B} Z_{\nu} r^{-1}_{i\nu} \tag{V.2.13}$$

is the (one-electron) potential energy operator of interaction between electrons of the molecule A and nuclei (nuclear charge Z_{ν}) of the molecule B and

$$\hat{V}^{AB}_{ee} = \sum_{i\in A} \sum_{j\in B} r^{-1}_{ij} \tag{V.3.14}$$

the analogous operator corresponding to the potential energy of interaction between electrons of the molecule A and electrons of the molecule B.

The repulsion energy between nuclear charges embedded in the molecules A and B is contained in $\hat{V}^{AB}_{nn}$; as clearly seen, it contributes nothing to the 2-nd order energy, so that it may be ignored.

Thanks to eqs. (V.3.10), (V.3.12) the expressions for the 2-nd order polarization and exchange polarization energy can be given the following form

$$\begin{aligned} E_{02} = {} & \langle\varphi^{A}_{0}|(\hat{V}^{A}_{en} + \langle\varphi^{B}_{0}|\hat{V}^{AB}_{ee}|\varphi^{B}_{0}\rangle)\hat{R}^{A}_{0}(\hat{V}^{A}_{en} + \langle\varphi^{B}_{0}|\hat{V}^{AB}_{ee}|\varphi^{B}_{0}\rangle)|\varphi^{A}_{0}\rangle \\ & + \langle\varphi^{B}_{0}|(\hat{V}^{B}_{en} + \langle\varphi^{A}_{0}|\hat{V}^{AB}_{ee}|\varphi^{A}_{0}\rangle)\hat{R}^{B}_{0}(\hat{V}^{B}_{en} + \langle\varphi^{A}_{0}|\hat{V}^{AB}_{ee}|\varphi^{A}_{0}\rangle)|\varphi^{B}_{0}\rangle \\ & + \langle\varphi^{A}_{0}\varphi^{B}_{0}|\hat{V}^{AB}_{ee}\hat{R}^{AB}_{0}\hat{V}^{AB}_{ee}|\varphi^{A}_{0}\varphi^{B}_{0}\rangle \end{aligned} \tag{V.3.15}$$

$$\begin{aligned} E_{02,ex} = {} & \langle\varphi^{A}_{0}|(\hat{V}^{A}_{en}+\langle\varphi^{B}_{0}|\hat{V}^{AB}_{ee}|\varphi^{B}_{0}\rangle)\hat{R}^{A}_{0}\langle\varphi^{B}_{0}|(\hat{\mathcal{P}}^{AB}-\mathcal{P}^{AB}_{0})(\hat{V}-E_{01})|\varphi^{A}_{0}\varphi^{B}_{0}\rangle \\ & + \langle\varphi^{B}_{0}|(\hat{V}^{B}_{en}+\langle\varphi^{A}_{0}|\hat{V}^{AB}_{ee}|\varphi^{A}_{0}\rangle)\hat{R}^{B}_{0}\langle\varphi^{A}_{0}|(\hat{\mathcal{P}}^{AB}-\mathcal{P}^{AB}_{0})(\hat{V}-E_{01})|\varphi^{A}_{0}\varphi^{B}_{0}\rangle \\ & + \langle\varphi^{A}_{0}\varphi^{B}_{0}|\hat{V}^{AB}_{ee}\hat{R}^{AB}_{0}(\hat{\mathcal{P}}^{AB}-\mathcal{P}^{AB}_{0})(\hat{V}-E_{01})|\varphi^{A}_{0}\varphi^{B}_{0}\rangle \end{aligned} \tag{V.3.16}$$

The quantites $<\varphi_0^A|\hat{V}_{ee}^{AB}|\varphi_0^A>$ and $<\varphi_0^B|\hat{V}_{ee}^{AB}|\varphi_0^B>$ appearing in the above expressions are <u>operators</u> which depend on the positions of the electrons belonging to the "monomers" B and A, respectively. $<\varphi_0^A|\hat{V}_{ee}^{AB}|\varphi_0^A>$ represents the electrostatic potential energy operator for the electrons belonging to the molecule B, in the <u>average</u> field produced by the <u>unperturbed</u> <u>electronic</u> <u>distribution</u> of the molecule A, so that $\hat{V}_{en}^A$ + $<\varphi_0^B|\hat{V}_{ee}^{AB}|\varphi_0^B>$ is the overall (i.e. nuclear plus electronic), effective, potential energy operator for the electrons in A due to the presence of the unperturbed molecule B.

The first matrix element in eq. (V.3.15),

$$E_{02,ind}^A = <\varphi_0^A|(\hat{V}_{en}^A + <\varphi_0^B|\hat{V}_{ee}^{AB}|\varphi_0^B>)\hat{R}_0^A(\hat{V}_{en}^A + <\varphi_0^B|\hat{V}_{ee}^{AB}|\varphi_0^B>)|\varphi_0^A> \qquad (V.3.17)$$

is the 2-nd order <u>induction</u> energy of the molecule A[2], a contribution to the 2-nd order polarization energy E_{02} associated with the distortion of the ground state charge distribution of that molecule in the <u>average</u> field produced by the other. There is, by symmetry reasons, an entirely analogous contribution $E_{02,ind}^B$ to E_{02}, which formally stems from eq. (V.3.17) through the exchange B⇄A (second matrix element on the right-hand side of eq. (V.3.15)).

The last contribution to E_{02},

$$E_{02,disp} = <\varphi_0^A\varphi_0^B|\hat{V}_{ee}^{AB}\hat{R}_0^{AB}\hat{V}_{ee}^{AB}|\varphi_0^A\varphi_0^B> \qquad (V.3.18)$$

is traditionally known as the <u>(2-nd order) dispersion energy</u> [2,4]. Contrary to the induction energy, which is related to a mean field effect, the dispersion energy is intimately associated with $\hat{V}_{ee}^{AB}$, the <u>local</u> inter-partner electron repulsion energy operator. For this reason, one speaks a bit loosely of the dispersion energy as an effect caused by coupling between "instantaneous" electronic motions in the two "monomers" [2] (actually no time is involved in our analysis).

In order to appreciate the origin of the dispersion energy in a different way, let us consider the explicit expression of $\hat{R}_0^{AB}$, eqs. (V.3.11), and write it in the approximate form, first proposed by Unsöld (see [2]-[4]),

$$\hat{R}_0^{AB} \simeq -(\overline{\Delta E_{AB}})^{-1}[\hat{1} - \hat{O}_0^A][\hat{1} - \hat{O}_0^B] \qquad (V.3.19)$$

$(\overline{\Delta E_{AB}})$ being a suitable average excitation energy. If eq. (V.3.19) is introduced into eq. (V.3.18), one finds very easily

$$E_{02,disp} \simeq -(\overline{\Delta E_{AB}})^{-1} \langle \varphi_0^A \varphi_0^B | \hat{V}_{ee}^{AB} (\hat{1} - \hat{\mathcal{O}}_0^A)(\hat{1} - \hat{\mathcal{O}}_0^B) \hat{V}_{ee}^{AB} | \varphi_0^A \varphi_0^B \rangle$$

$$= -(\overline{\Delta E_{AB}})^{-1} \langle \varphi_0^A \varphi_0^B | [\hat{V}_{ee}^{AB} - \langle \varphi_0^A | \hat{V}_{ee}^{AB} | \varphi_0^A \rangle][\hat{V}_{ee}^{AB} - \langle \varphi_0^B | \hat{V}_{ee}^{AB} | \varphi_0^B \rangle] | \varphi_0^A \varphi_0^B \rangle$$

The operator $\delta \hat{V}_{ee}^{AB} = -[\hat{V}_{ee}^{AB} - \langle \varphi_0^A | \hat{V}_{ee}^{AB} | \varphi_0^A \rangle]$ can be interpreted in a straightforward way as the deviation or <u>fluctuation</u> of the potential (operator) felt by the electrons belonging to the molecule B, due to the electrons of the molecule A, from its mean value $-\langle \varphi_0^A | \hat{V}_{ee}^{AB} | \varphi_0^A \rangle$ (see the remarks following eq. (V.3.16)), with an entirely equivalent interpretation for $-[\hat{V}_{ee}^{AB} - \langle \varphi_0^B | \hat{V}_{ee}^{AB} | \varphi_0^B \rangle]$. One says therefore that the dispersion energy is caused by the correlation of "fluctuations" of the potential felt by the electrons in both molecules.

As clearly seen from eqs. (V.3.17), (V.3.18), induction and dispersion energies are <u>negative</u>-definite quantities for molecule <u>both</u> in their <u>ground state</u>, that means an <u>attractive</u> contribution to the interaction energy as far as the polarization energy E_{02} is concerned.

The exchange polarization energy $E_{02,ex}$ can also be partitioned in induction and dispersion components; from eq. (V.3.16), we may write

$$E_{02,ex\text{-}ind} = \langle \varphi_0^A | (\hat{V}_{en}^A + \langle \varphi_0^B | \hat{V}_{ee}^{AB} | \varphi_0^B \rangle) \hat{R}_0^A \langle \varphi_0^B | (\hat{\mathcal{P}}^{AB} - \mathcal{P}_0^{AB})$$

$$\cdot (\hat{V} - E_{01}) | \varphi_0^A \varphi_0^B \rangle + \langle \varphi_0^B | (\hat{V}_{en}^B + \langle \varphi_0^A | \hat{V}_{ee}^{AB} | \varphi_0^A \rangle) \hat{R}_0^B \langle \varphi_0^A | (\hat{\mathcal{P}}^{AB} - \mathcal{P}_0^{AB})$$

$$\cdot (V - E_{01}) | \varphi_0^A \varphi_0^B \rangle \qquad (V.3.20)$$

$$E_{02,ex\text{-}disp} = \langle \varphi_0^A \varphi_0^B | \hat{V}^{AB} \hat{R}_0^{AB} (\hat{\mathcal{P}}^{AB} - \mathcal{P}_0^{AB}) (V - E_{01}) | \varphi_0^A \varphi_0^B \rangle \qquad (V.3.21)$$

for the <u>exchange induction energy</u> and the <u>exchange dispersion energy</u> respectively. The quantities $\langle \varphi_0^A | (\hat{\mathcal{P}}^{AB} - \mathcal{P}_0^{AB})(\hat{V} - E_{01}) | \varphi_0^A \varphi_0^B \rangle$ and $\langle \varphi_0^B | (\hat{\mathcal{P}}^{AB} - \mathcal{P}_0^{AB})(\hat{V} - E_{01}) | \varphi_0^A \varphi_0^B \rangle$ are to be regarded as operators defined in the Hilbert spaces spanned by $\{|\varphi_m^B\rangle\}$ and $\{|\varphi_n^A\rangle\}$, respectively.

Coming back to eqs. (V.3.8) and (V.3.9) and making use of the partitions expressed by eqs. (V.3.15) and (5.3.16), the interaction energy ΔE through the 2-nd order, in the region of small overlap values, can be cast in the following form

$$\Delta E = (E_{01}+E_{02,ind}+E_{02,disp}) + (E_{01,ex}+E_{02,ex\text{-}ind}+E_{02,ex\text{-}disp}) \quad (V.3.21)$$

It should be clear that the quantity within the first parentheses is exactly the energy through the 2-nd order that would arise from the simple "polarization approximation" approach, so that $(E_{01,ex}+E_{02,ex\text{-}ind}+E_{02,ex\text{-}disp})$ is the "surplus" from using a symmetry-adapted perturbation theory; this latter contribution is generally dependent on the chosen symmetry-adapted formalism.

V.4. Many-Electron Systems: the Problems Posed by Our Ignorance of Exact Unperturbed Eigenstates.

The straightforward application of the perturbation formalisms reviewed in the past sections to more complicate situations than $H...H^+$ is actually limited to interacting one-electron atoms (also in this case, at the cost of heavy difficulties, connected with the practical manipulation of the Rayleigh-Schrödinger resolvent $\hat{R}_0$). If we want to extend the approach to interactions involving many-electron atoms and, possibly, molecules , some changes, not exactly trivial, are needed, so as to take into account that we do not dispose of exact eigenstates to the unperturbed Hamiltonian operator $\hat{H}_0$, not even for the very ground state, our equipment as a rule consisting of approximate wavefunctions for a severely reduced number of states.

A possible way out from these complications relies on the recognition that approximate wavefunctions of the Hartree-Fock (H.F.) type [7,33,106] usually provide a good description for the ground state of atoms and molecules, and systematic procedures can be set up to correct for the correlation effects [106-109] missing in such approximation. Even though this statement does not imply a lack of problems, it is a fact that one may look optimistically at the present-day computational facilities, by which not only the production of accurate self-consistent field (SCF) wavefunctions has become a fairly routine procedure, but the overcoming of such approximation level, for example along perturbative lines, by now appears a promising possibility [106,109,110].

We now propose to dwell upon the way we may proceed for a perturbative approach either to the unperturbed ground state solution $|\varphi_0\rangle$ to the Hamiltonian $\hat{H}_0$ or the resolvent $\hat{R}_0$, two basic ingredients of any realistic calculation of intermolecular interactions.

Speaking in general, let us assume to have at our disposal an orthonormal set $\{|\chi_i\rangle\}$ of one-electron <u>spinorbitals</u>, eigenstates to some one-electron Hamiltonian $\hat{h}$

$$\hat{h}(1)|\chi_i(1)\rangle = \varepsilon_i|\chi_i(1)\rangle \qquad (V.4.1)$$

where 1 denotes a particular electron. From the N eigenstates $|\chi_i\rangle$ associated with the lowest eigenvalues ε_i we may now build up an antisymmetric N - electron eigenstate, in the form of (normalized) Slater determinant,

$$|\Phi_0\rangle = (N!)^{\frac{1}{2}}\hat{A}||\chi_1(1)\rangle|\chi_2(2)\rangle\ldots|\chi_N(N)\rangle| \qquad (V.4.2)$$

where $\hat{A}$ is the antisymmetrizer [see eq. (III.3.2)]. It is easily verified that $|\Phi_0\rangle$ is eigensolution to the separable Hamiltonian $\hat{F}_0$,

$$\hat{F}_0 = \sum_{i=1}^{N} \hat{h}(i) \qquad (V.4.3)$$

with energy

$$W_0 = \sum_{i=1}^{N} \varepsilon_i \qquad (V.4.4)$$

i.e.

$$\hat{F}_0|\Phi_0\rangle = W_0|\Phi_0\rangle \qquad (V.4.5)$$

Beside $|\Phi_0\rangle$ we are in a position to generate excited eigenstates to $\hat{F}_0$, orthogonal to $|\Phi_0\rangle$ and to each other, by simply replacing one or more occupied spinorbitals appearing in $|\Phi_0\rangle$ by <u>unoccupied</u> $|\chi_i\rangle$'s belonging to the set $\{|\chi_i\rangle\}$. If we use Greek letters $\alpha,\beta,\gamma,\ldots.$ to represent spinorbitals occupied in $|\Phi_0\rangle$ and latin letters $r,s,t,\ldots.$ to represent the unoccupied spinorbitals, $\{|\Phi_\alpha^r\rangle\}$ is the totality of the single-excited configurations where one electron has jumped from the occupied spinorbital α into the r-th unoccupied, $\{|\Phi_{\alpha\beta}^{rs}>\}$ the set of all doubly-excited configurations, etc. These excited configurations are eigensolutions to $\hat{F}_0$,

$$\hat{F}_0|\Phi_k> = W_k|\Phi_k> \tag{V.4.6}$$

where $|\Phi_k>$ and W_k stand for $|\Phi_{\alpha\beta..}^{rs..}>$ and $W_{\alpha\beta..}^{rs..}$, respectively, and clearly

$$W_{\alpha\beta..}^{rs..} = W_0 + (\varepsilon_r+\varepsilon_s+..)-(\varepsilon_\alpha+\varepsilon_\beta+..) \tag{V.4.7}$$

If $|\Phi_0>$ denotes our reference state, $\{|\Phi_\alpha^r>\}\oplus\{|\Phi_{\alpha\beta}^{rs}>\}\oplus...$ is a set spanning the orthogonal complement to $|\Phi_0>$.

The perturbation operator $\hat{U}$ is defined as

$$\hat{U} = \hat{H}_0 - \hat{F}_0 \tag{V.4.8}$$

and its structure obviously depends on the choice of $\hat{h}(i)$ [107,111].

If $\hat{T}(i)$ and $\hat{V}_{en}(i)$ represent respectively the kinetic energy operator and the potential energy operator of the electron i in the field of the nuclei only,

$$\hat{h}^{BN}(i) = \hat{T}(i) + \hat{V}_{en}(i)$$

is a possible candidate; in this case $\{|\chi_i>\}$ is a set of "bare nuclei" eigenstates and the perturbation includes only inter-electronic repulsion terms. If screened nuclear charges are used in $\hat{V}_{en}(i)$, our basis $\{|\chi_i>\}$ will consist of "screened nuclei" eigenstates, and the perturbation $\hat{U}$ now includes both inter-electronic and electron-nucleus interaction terms. A particularly important choice corresponds to the case where $\hat{h}(i)$ is the <u>Hartree-Fock</u> <u>effective</u> <u>Hamiltonian</u>, i.e.

$$\hat{h}^{HF}(i) = \hat{T}(i)+\hat{V}_{en}(i)+\hat{V}_{HF}^N(i) \equiv \hat{h}^{BN}(i) + \hat{V}_{HF}^N(i) \tag{V.4.9}$$

where $\hat{V}_{HF}^N(i)$ is the <u>"unrestricted"</u> <u>Hartree-Fock</u> <u>potential</u> <u>energy</u> <u>operator</u> [7,33,107], by which the actual inter-electron repulsion effects are approximately taken into account in a self-consistent manner. As well known, such operator is so defined,

$$\hat{V}^N_{HF}(\underline{x}_1)\chi_i(\underline{x}_1) = \sum_{\alpha=1}^{N} |\int d\underline{x}_2 |\chi_\alpha(\underline{x}_2)|^2 r_{12}^{-1}\chi_i(\underline{x}_1) \quad \text{direct or Coulomb term}$$

$$- \int d\underline{x}_2 \chi_\alpha^*(\underline{x}_2)\chi_i(\underline{x}_2) r_{12}^{-1}\chi_\alpha(\underline{x}_1)| \quad \text{exchange term} \qquad (V.4.10)$$

$\underline{x}_i \equiv (\underline{r}_i,\sigma_i)$ denoting the space-spin coordinates of the i-th electron. From eq. (V.4.10) it appears that the occupied one-electron states $(i < N)$ are evaluated in the field of the nuclei and N-1 electrons, because direct and exchange terms cancel each other, while no cancellation of this kind occurs for unoccupied states, which are therefore evaluated in the field of the nuclei and N electrons. This fact bears consequences: it is known, in fact, that the occupied eigenstates to $\hat{h}^{HF}$ correspond to bound states, while for neutral systems all unoccupied states lie in the continuum [112,113].

In view of these spectral features of $\hat{h}^{HF}$, which hint at a possible slowly convergent behaviour of expansions founded on the set of H.F. spinorbitals, it has been suggested as advantageous in perturbation theory the use of a Hamiltonian $\hat{h}$ characterized by a potential in which the one-electron excited states correspond more closely to true single excitations in the atom [112,113]. This program can be attained by using a (N-1) electron effective potential $\hat{V}^{N-1}$ that provides the same occupied spinorbitals as $\hat{V}^N_{HF}$ and unoccupied spinorbitals with partly discrete and partly continuum eigenvalues; thus, we may write

$$\hat{h}(i) = \hat{T}(i) + \hat{V}_{en}(i) + \hat{V}^{N-1}(i) \qquad (V.4.11)$$

A rather favoured choice for $\hat{V}^{N-1}$ is the following [114,115]

$$\hat{V}^{N-1} = \hat{V}^N_{HF} + \hat{P}\hat{\Omega}\hat{P} \qquad (V.4.12)$$

where $\hat{P} = \hat{1} - \sum_{\alpha=1}^{N}|\chi_\alpha\rangle\langle\chi_\alpha|$ is the projector on the orthogonal complement to the subspace spanned by the N occupied spinorbitals and $\hat{\Omega}$ an arbitrary (one-electron) self-adjoint operator. Obviously, $\hat{V}^{N-1}|\chi_\beta\rangle = \hat{V}^N_{HF}|\chi_\beta\rangle$ and $\hat{V}^{N-1}|\chi_r\rangle = [\hat{V}^N_{HF} + \hat{P}\hat{\Omega}]|\chi_r\rangle$, so that through a suitable choice of $\hat{\Omega}$, the desired (N-1)-electron shielding also for unoccupied eigenstates becomes possible. The corresponding Hamiltonian $\hat{h}^{KSH}$ of eq. (V.3.11) can be referred to as Kelly-Silverstone-Huzinaga Hamiltonian.

The perturbation theory founded on the unperturbed Hamiltonian $\hat{F}_0 = \sum_{i=1}^{N} \hat{h}^{HF}(i)$ [or $\hat{F}_0 = \sum_{i=1}^{N} \hat{h}^{KSH}(i)$] is known as "unrestricted" Møller - Plesset procedure [110,116] and has extensively been considered from the diagrammatic point of view [109,117] (a large quantity of references will be found in these two papers). However, it is not the only one. Another approach, frequently referred to as Epstein-Nesbet procedure [118-120], is based on the following unperturbed Hamiltonian

$$\hat{F}_0^{EN} = \sum_k |\Phi_k\rangle\langle\Phi_k|\hat{H}|\Phi_k\rangle\langle\Phi_k| \tag{V.4.13}$$

where $\{|\Phi_k\rangle\}$ is still the set of Slater determinants obtained from the set $\{|\chi_i\rangle\}$. Clearly,

$$\hat{F}_0^{EN}|\Phi_k\rangle = |\Phi_k\rangle\langle\Phi_k|\hat{H}|\Phi_k\rangle \tag{V.4.14}$$

and the eigenvalues do $\hat{F}_0^{EN}$ are now the diagonal matrix elements of the full Hamiltonian $\hat{H}$ in the basis $\{|\Phi_k\rangle\}$; in particular, the energy value associated with the unperturbed ground state $|\Phi_0\rangle$ is the H.F. energy $\langle\Phi_0|\hat{H}|\Phi_0\rangle$, if the set $\{|\chi_i\rangle\}$ consists of one-electron solutions to $\hat{h}^{HF}$ or $\hat{h}^{KSH}$. The perturbation in this case is therefore essentially caused by the off-diagonal matrix elements $\langle\Phi_k|\hat{H}|\Phi_l\rangle$ on the diagonal ones.

In what follows we shall essentially be concerned with ground states corresponding to <u>closed electronic shells</u>. In these cases, contrary to the general situation, the H.F. Hamiltonian $\hat{F}_0$ surely has the full symmetry of the molecular framework and does not depend upon the spin; thus, $|\Phi_0\rangle$ corresponds to a pure spin state and is not degenerate, with paired spinorbitals of the form

$$\begin{aligned} |\chi_{2i-1}\rangle &\equiv |\psi_i\rangle|+\tfrac{1}{2}\rangle \\ |\chi_{2i}\rangle &\equiv |\psi_i\rangle|-\tfrac{1}{2}\rangle \end{aligned} \qquad (i=1,2,\ldots,N/2) \tag{V.4.15}$$

where $|\psi_i\rangle$ denotes an orbital and $|\pm\frac{1}{2}\rangle$ are spin eigenstates. Since these spinorbitals are solutions to the "unrestricted" H.F. equations, i.e.

$$\hat{h}^{HF}|\chi_i\rangle = \varepsilon_i|\chi_i\rangle \tag{V.4.16}$$

the formal simplicity of the "unrestricted" theory is retained, along with some important results, particularly the Brillouin's theorem [7, 33]. As a justification of our interest in closed-shell situations, we only remind that the ground state of nearly all stable molecules falls within this class.

The development of a perturbation procedure for the exact eigenstate $|\varphi_0\rangle$ to $\hat{H}_0$ and the associated exact eigenvalue E_0 can obviously be founded on the general partitioning technique reviewed in Chap.III. Thus, from eq. (III.2.8) we get

$$|\varphi_0\rangle = |\Phi_0\rangle + \hat{T}(E_0)\hat{U}|\Phi_0\rangle \tag{V.4.17}$$

where to the "intramolecular" perturbation operator $\hat{U}$, defined by eq. (V.3.8), is entrusted the task of introducing the correlation effects missing in $|\Phi_0\rangle$; according to eq. (III.1.5) the reduced resolvent $\hat{T}(E_0)$ is given by

$$\hat{T}(E_0) = [E_0 - \hat{P}_0\hat{H}_0\hat{P}_0]^{-1}\hat{P}_0 \tag{V.4.18}$$

$\hat{P}_0$ being the projector onto the orthogonal complement to $|\Phi_0\rangle$:

$$\hat{P}_0 = \hat{1} - |\Phi_0\rangle\langle\Phi_0| \tag{V.4.19}$$

Analogously, assuming that $\langle\Phi_0|\Phi_0\rangle = 1$, the eigenvalue E_0 is related to W_0 through eq. (III.1.11)

$$E_0 = W_0 + \langle\Phi_0|\hat{U}|\Phi_0\rangle + \langle\Phi_0|\hat{U}\hat{T}(E_0)\hat{U}|\Phi_0\rangle \tag{V.4.20}$$

An equivalent expression for the reduced resolvent, which is suitable for iteration developments, is readily obtained from eq. (V.4.18). If we proceed along lines similar to those used for $\hat{T}_1(\varepsilon)$ in sec. III.3, we may write $\hat{T}(E_0) = [(W_0 - \hat{P}_0\hat{F}_0) - \hat{P}_0\hat{U}']^{-1}\hat{P}_0$, where

$$\hat{U}' = \hat{U} - (E_0 - W_0) \qquad (V.4.21)$$

and some manipulations then lead to the following expression for $\hat{T}(E_0)$ which is similar to $\hat{T}_1(\varepsilon)$ [eq. (III.3.10)],

$$\hat{T}(E_0) = [\hat{1} - \hat{R}_{00}\hat{U}']^{-1}\hat{R}_{00} \qquad (V.4.22)$$

where

$$\hat{R}_{00} = [W_0 - \hat{F}_0]^{-1}\hat{P}_0 \qquad (V.4.23)$$

is the relevant Rayleigh-Schrödinger resolvent.

From eqs. (V.4.17) and (V.4.22), the lowest order correction to the H.F. approximation which takes into account correlation effects results

$$|\Phi_{01}\rangle = \hat{R}_{00}\hat{U}|\Phi_0\rangle = \hat{R}_{00}\hat{H}_0|\Phi_0\rangle \qquad (V.4.24)$$

and we recover the well known fact that (at such order) only two-excited configurations of the form $|\Phi^{rs}_{\alpha\beta}\rangle$ are to be retained in the spectral resolution of $\hat{R}_{00}$, as a consequence of the Brillouin's theorem.

The Rayleigh-Schrödinger resolvent $\hat{R}_0 = [E_0 - \hat{H}_0]^{-1}[\hat{1} - |\varphi_0\rangle\langle\varphi_0|] = [E_0 - \hat{H}_0]^{-1}\hat{P}$ (not to be confused with $\hat{R}_{00}$ or $\hat{T}(E_0)$) is another basic quantity for the evaluation of interaction energies beyond the 1-st order, so that an expression for this operator in terms of intramolecular correlation corrections appears of interest.

The projector $\hat{P} = \hat{1} - |\varphi_0\rangle\langle\varphi_0|$ is readily expressed in terms of a perturbation series involving $\hat{U}$; from eqs. (V.4.17) and (V.4.22), we get through the 1-st order in $\hat{U}$

$$\hat{P} = \hat{P}_0 + \hat{R}_{00}\hat{U}\hat{P}_0 + \hat{P}_0\hat{U}\hat{R}_{00} - (\hat{R}_{00}\hat{U} + \hat{U}\hat{R}_{00}) + O(\hat{U}^2) \qquad (V.4.25)$$

Moreover, the very definition of $\hat{R}_0$

$$\hat{R}_0 = [E_0 - \hat{H}_0]^{-1}\hat{P} = [(W_0 - \hat{F}_0) - \hat{U}']^{-1}\hat{P} = [\hat{1} - (W_0 - \hat{F}_0)^{-1}\hat{U}']^{-1}(W_0 - \hat{F}_0)^{-1}\hat{P}$$

and the use of the general identity eq. (III.3.14) along with eq. (V.

4.25) lead in a straightforward way to the searched result

$$\hat{R}_0 = \hat{R}_{00}+\hat{R}_{00}(\hat{U}-W_{01})\hat{R}_{00}-\frac{1}{2}[\hat{R}^2_{00}\hat{U}(\hat{1}-\hat{P}_0)+(\hat{1}-\hat{P}_0)\hat{U}\hat{R}^2_{00}]+O(\hat{U}^2) \qquad (V.4.26)$$

which is correct through the 1-st order in $\hat{U}$. We have put $W_{01} = \langle\Phi_0|\hat{U}|\Phi_0\rangle$, a quantity clearly missing if the Epstein-Nesbet partition of $\hat{H}_0$ is employed.

Although the procedure now described is surely suitable for introducing intra-correlation effects in the calculation of intermolecular interaction energies, its actual application even at the lowest orders is bound to meet with noticeable difficulties, except possibly for the case of simple systems [121,122]. Intra-correlation effects have more frequently been either entirely neglected or approximately taken into account without any reference to a definite intra-molecular perturbation order. The first case formally corresponds to use in the relevant inter-molecular perturbation equations the model Hamiltonian operator $\hat{F}_0 = \sum_i \hat{h}^{HF}(i)$ in place of $\hat{H}_0$; the resulting formalism is known as uncoupled H.F. perturbation theory [2,19,123,124] and leads to generally unsatisfactory results for induction as well as dispersion energy contributions, in line with the poor quality of 2-nd order molecular properties (for example, electric polarizability) obtained by this method [125]. Different choiches of the one-electron operator $\hat{h}(i)$ from $\hat{h}^{HF}(i)$ are expected to lead to sensitive differences with respect to the uncoupled H.F. results: we shall see some specific examples of this behaviour in the next chapters. In the second case, intra-correlation effects have often been introduced by making recourse to the so-called coupled H.F. perturbation theory, [2,19,123,124], an approach rather commonly employed for the evaluation of 2-nd order atomic (molecular) properties, that is shown to be correct through the 1-st order in the electron correlation (it partially includes also higher-order contributions) [126]: in the course of Chap. VII we shall go into some details, but it is convenient to point out by now that a still better appreciation of the intra-correlation role may be necessary for sensible results to be obtained (see for instance ref. [127]).

The hard problem posed by the necessity of taking into account correlation effects in the evaluation of intermolecular interactions are not at all peculiar of the perturbative approach, but dö appear as a critical step also in variational calculations. The correlation energy error which affects the "monomer" unperturbed energies is balanced only in a very rough way by the analogous error made in evaluating the "complex" energy, so that a net correlation energy contribution to the interaction energy has to be taken into account. Since the interaction energy ΔE over a wide range of distances is often a small fraction of the absolute error by which the calculated correlation energies are affected and this error fluctuates as the relative distance between the "monomers" changes [5], meaningful results for ΔE are obtained only at the price of very accurately balanced calculations. In particular, what can possibly be mere "basis superposition" or "ghost orbital" effects [128,130], i.e. spurious contributions to the interaction energy arising from the use of unsaturated basis sets for each "monomer", should be carefully discerned or avoided.

VI. THE CALCULATION OF THE 1-ST ORDER INTERACTION ENERGY.

VI.1. Coulombic Contribution to the 1-st Order Interaction Energy: a Useful Expression in Terms of the Charge Density Matrix.

In order to pave the way for further developments, we begin this section by writing down an equivalent expression for the Coulombic portion E_{01} of the 1-st order perturbation energy, eq. (V.3.8), which is of value for both practical and conceptual reasons.

To this end, let us introduce the electric charge density operator $\hat{\rho}_A(\underline{r})$ for the molecule A,

$$\hat{\rho}_A(\underline{r}) = \hat{\rho}_A^{nuc}(\underline{r}) + \hat{\rho}_A^{el}(\underline{r}) \qquad \text{(VI.1.1)}$$

$$\hat{\rho}_A^{nuc}(\underline{r}) = \sum_{\mu \in A} Z_\mu \delta(\underline{r} - \underline{R}_\mu) \; , \qquad \hat{\rho}_A^{el}(\underline{r}) = - \sum_{i \in A} \delta(\underline{r} - \underline{r}_i)$$

where Z_μ denotes the electric charge (in a.u.) of the μ-th nucleus located at the position $\underline{R}_\mu$ in the molecule A, while $\underline{r}_i$ is the position vector of the i-th electron in the same molecule.* An entirely analogous expression can obviously be written for the electric charge density operator $\hat{\rho}_B(\underline{r})$ of the molecule B.

It is now a simple exercise to verify that the interaction potential energy operator $\hat{V}$ for two interacting "monomers" [see eqs. (V.3.12) - (V.3.14)],

$$\hat{V} = \hat{V}_{en}^{A} + \hat{V}_{en}^{B} + \hat{V}_{ee}^{AB} + \hat{V}_{nn}^{AB} =$$

$$= - \sum_{i \in A} \sum_{\nu \in B} Z_\nu r_{i\nu}^{-1} - \sum_{j \in B} \sum_{\mu \in A} Z_\mu r_{j\mu}^{-1} + \qquad \text{(VI.1.2)}$$

$$+ \sum_{i \in A} \sum_{j \in B} r_{ij}^{-1} + \sum_{\mu \in A} \sum_{\nu \in B} Z_\mu Z_\nu R_{\mu\nu}^{-1}$$

can compactly be written in the suggestive integral form [131,132]

---

*We should actually regard $\underline{r}_i$ as an operator quantity (to be written $\hat{\underline{r}}_i$, so as to fit in with our general notation); we are, however, tacitly assuming that the coordinate picture is going to be used.

$$\hat{V} = \iint d\underline{r}\, d\underline{r}' \frac{\hat{\rho}_A(\underline{r})\ \hat{\rho}_B(\underline{r}')}{|\underline{r} - \underline{r}'|} \qquad \text{(VI.1.3)}$$

reminiscent of the classical expression for the interaction energy between two continuous charge distributions.

If we introduce this expression for $\hat{V}$ in the standard matrix element

$$E_{01} = \langle \varphi_o^A\ \varphi_o^B\ |\hat{V}|\ \varphi_o^A\ \varphi_o^B \rangle \qquad \text{(VI.1.4)}$$

[eq. (V.3.8)], we get the following alternative form for the Coulombic portion of the 1-st order interaction energy:

$$E_{01} = \iint d\underline{r}\, d\underline{r}' \frac{\langle \hat{\rho}_A(\underline{r})\rangle_o \langle \hat{\rho}_B(\underline{r}')\rangle_o}{|\underline{r}' - \underline{r}|} \qquad \text{(VI.1.5)}$$

where

$$\langle \hat{\rho}_A(\underline{r})\rangle_o = \langle \varphi_o^A\ |\hat{\rho}_A(\underline{r})|\ \varphi_o^A \rangle \qquad \text{(VI.1.6)}$$

is the mean value, at the point $\underline{r}$, of the unperturbed electric charge density of the molecule A in the state $|\varphi_o^A\rangle$.

Eq. (VI.1.5) provides in a transparent way the reason for the appelation of electrostatic energy reserved to E_{01}; the result is formally a purely classical one, the same that one would obtain for two smeared static electric charge distributions [133], the quantum-mechanical vestige being only contained in the way the involved charge densities are to be calculated [eq. (VI.1.6)].

The charge density value $\langle \hat{\rho}(\underline{r})\rangle_o$ at any point $\underline{r}$ can be given a compact form in terms of the first-order density matrix [7,134-136]; we have, in fact,

$$\langle \hat{\rho}_A(\underline{r})\rangle_o \equiv \langle \varphi_o^A|\hat{\rho}_A(\underline{r})|\varphi_o^A\rangle = \sum_{\mu\in A} Z_\mu \delta(\underline{r} - \underline{R}_\mu) - \int d\underline{x}_1 \ldots d\underline{x}_{N^A}\ \varphi_o^{A*}(\underline{x}_1 \ldots \underline{x}_{N^A})$$

$$\cdot \sum_{i=1}^{N^A} \delta(\underline{r} - \underline{r}_i)\, \varphi_o^A(\underline{x}_1 \ldots \underline{x}_{N^A}) = \sum_{\mu\in A} Z_\mu \delta(\underline{r} - \underline{R}_\mu)$$

$$- N^A \int d\underline{x}_1 \delta(\underline{r} - \underline{r}_1) \int \ldots \int d\underline{x}_2 \ldots d\underline{x}_{N^A} \varphi_o^{A*}(\underline{x}_1 \ldots \underline{x}_{N^A})$$

$$\cdot \varphi_0^A(\underline{x}_1 \ldots \underline{x}_{N^A}) = \sum_{\mu \in A} Z_\mu \delta(\underline{r} - \underline{R}_\mu)$$

$$- \int d\underline{x}_1 \delta(\underline{r} - \underline{r}_1) \rho_A(\underline{x}_1, \underline{x}_1) \qquad \text{(VI.1.7)}$$

where we have introduced the first-order density matrix $\rho_A(\underline{x}_1, \underline{x}_1')$

$$\rho_A(\underline{x}_1, \underline{x}_1) = N^A \int \ldots \int d\underline{x}_2 \ldots d\underline{x}_{N^A} \varphi_0^{A*}(\underline{x}_1 \ldots \underline{x}_{N^A}) \varphi_0^A(\underline{x}_1 \ldots \underline{x}_{N^A}) \qquad \text{(VI.1.8)}$$

In the formulae above, as usual $\underline{x} \equiv (\underline{r}, \sigma)$ and the integration with respect to d$\underline{x}$ actually means integration over the configuration space (d$\underline{r}$) and summation over the two allowed spin eigenvalues.

The result expressed by eq. (VI.1.7) is quite rigorous, but its practical utility requires that we specialize it, taking into account that we actually dispose as a rule of approximate energy eigenstates (see sec. V.3).

Among these, the SCF wavefunctions constitute a very important class of accessible and reasonably accurate approximate energy eigenstates, so that an expression for the corresponding first-order density matrix $\rho_A(\underline{x}, \underline{x}')$ appears valuable. For a single Slater determinant wavefunction, built up in terms of N^A orthonormal spinorbitals $\chi_k(\underline{x})$, it can be shown that [134-136]

$$\rho_A(\underline{x}, \underline{x}') = \sum_{k=1}^{N^A} \chi_k(\underline{x}) \chi_k^*(\underline{x}') \qquad \text{(VI.1.9)}$$

The very important case of a closed-shell state, which we know to correspond to the situation where there are $N^A/2$ doubly occupied orbitals $\psi_k(\underline{r})$, is clearly contained in the more general result (VI.1.9).

As an explicit example, we may readily obtain the overall Coulomb interaction energy E_{01} in the case of two closed-shell molecular charge distributions A, B, approximately described by $N^A/2$ molecular orbitals $\{\psi_k^A\}$ belonging to A and $N^B/2$ molecular orbitals $\{\psi_j^B\}$ belonging to B, respectively (see Fig. VI.1); from eqs. (VI.1.5), (VI.1.6) and (VI.1.9) it follows

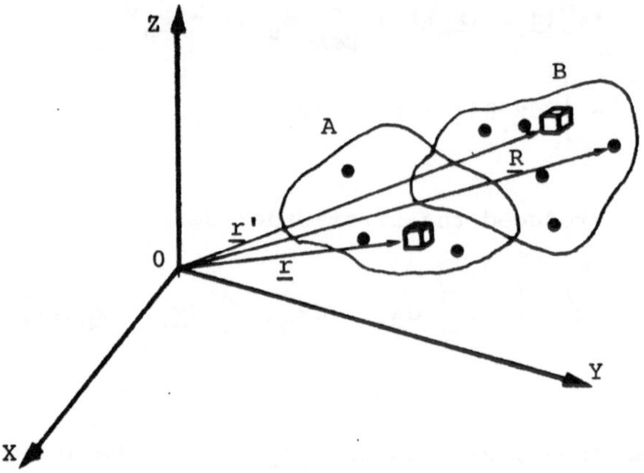

Fig. VI.1. Schematic view of two interacting electric charge distributions. The dots denote nuclear charges.

$$E_{01} = \sum_{\mu\in A}\sum_{\nu\in B} \frac{Z_\mu Z_\nu}{R_{\mu\nu}} - 2\sum_{\mu\in A}\sum_{j\in B} Z_\mu \int d\underline{r}' \frac{\left|\psi_j^B(\underline{r}')\right|^2}{\left|\underline{r}' - \underline{R}_\mu\right|}$$

$$- 2\sum_{\nu\in B}\sum_{i\in A} Z_\nu \int d\underline{r} \frac{\left|\psi_i^A(\underline{r})\right|^2}{\left|\underline{r} - \underline{R}_\nu\right|} \qquad \text{(VI.1.10)}$$

$$+ 4\sum_{i\in A}\sum_{j\in B} \int\int d\underline{r}\, d\underline{r}' \frac{\left|\psi_i^A(\underline{r})\right|^2 \left|\psi_j^B(\underline{r}')\right|^2}{\left|\underline{r} - \underline{r}'\right|}$$

The first term on the second-hand side of eq. (VI.1.10) is the nuclear repulsion energy between the two molecules, the second (third) one the attraction energy between nuclei of the molecule A(B) and the electronic distribution of the molecule B(A), while the fourth term corresponds to the repulsion energy between the electron "clouds" of the two molecules.

VI.2. The Evaluation of the Coulombic Energy.

Although eq. (VI.1.10) is transparent from the interpretative point of view, as it stands it is not necessarily a good starting point for computational purposes, particularly when the two charge distributions are far apart, because of the occurrence of noticeable cancellations between separately small contributions. The same conclusion is valid when more general forms of the first-order density matrix, eq. (VI.1.8), are used.

A useful approach to the evaluation of E_{01}, which does not incur the above mentioned difficulties, will now be put forward for the case of spherical atoms. To start with, let us consider the electron charge density operator $\hat{\rho}(\underline{r}) = -\sum_{i=1}^{N} \delta(\underline{r}-\underline{r}_i)$, one of the two contributions to the total charge density operator defined in eq. (VI.1.1). The following result is then immediately obtained for any neutral atom

$$\int d\underline{r} \; \langle \hat{\rho}^{el}(\underline{r}) \rangle_0 = -N \equiv -\mathfrak{Z} \qquad \text{(VI.2.1)}$$

$\mathfrak{Z}$ being the nuclear charge. Taking into account eq. (VI.2.1), the Coulomb interaction energy E_{01} can be expressed in terms of the electron charge density $\langle \hat{\rho}^{el}(\underline{r}) \rangle_0$ as follows

$$E_{01} = \int d\underline{r}_1 d\underline{r}_2 \langle \hat{\rho}_A^{el}(r_1) \rangle_0 \langle \hat{\rho}_B^{el}(r_2) \rangle_0 \left[\frac{1}{R} + \frac{1}{r_{12}} - \frac{1}{r_{1B}} - \frac{1}{r_{2A}}\right] \qquad \text{(VI.2.2)}$$

where, for convenience, we have labelled $\underline{r}_1$, $\underline{r}_2$ the position vectors previously denoted $\underline{r}$, $\underline{r}'$. We have, moreover, explicitly put in evidence that the electron charge densities are supposed spherically symmetric.

A convenient manipulation of eq. (VI.2.2) involves as a first step a four-fold integration with respect to the angular variables $\underline{\omega}_1 \equiv (\theta_1, \varphi_1)$, $\underline{\omega}_2 \equiv (\theta_2, \varphi_2)$, followed by the radial integrations:

$$E_{01}(R) = \int_0^\infty dr_1 \; 4\pi r_1^2 \langle \hat{\rho}_A^{el}(r_1) \rangle_0 \int_0^\infty dr_2 \; 4\pi r_2^2 \langle \hat{\rho}_B^{el}(r_2) \rangle_0 \, \mathfrak{J}(r_1, r_2; R) \qquad \text{(VI.2.3)}$$

where

$$\mathfrak{J}(r_1, r_2; R) = (4\pi)^{-2} \int d\underline{\omega}_1 \int d\underline{\omega}_2 \left[\frac{1}{R} + \frac{1}{r_{12}} - \frac{1}{r_{1B}} - \frac{1}{r_{2A}}\right] \qquad \text{(VI.2.4)}$$

The integrations involved in eq. (VI.2.4) are not particularly binding

[137,138], the result being expressible as

$$\mathfrak{I}(r_1, r_2; R) = \frac{1}{R} - 2\left[\frac{1}{|R+r_1+|R-r_1||}\right] + \left[\frac{1}{|R+r_2+|R-r_2||}\right] + F(r_1, r_2; R) \tag{VI.2.5}$$

where

$$F(r_1, r_2; R) \equiv (4\pi)^{-2}\int d\underline{\omega}_1 \int d\underline{\omega}_2 \frac{1}{r_{12}} = \begin{cases} \dfrac{2}{R+r_1+|R-r_1|}, & r_2 < |R-r_1| \\ \dfrac{1}{r_2}, & r_2 > R+r_1 \\ \dfrac{1}{2}\left(\dfrac{1}{r_1} + \dfrac{1}{r_2}\right) - \dfrac{R}{4r_1 r_2} - \dfrac{(r_1-r_2)^2}{4Rr_1 r_2}, & |R-r_1| < r_2 < R+r_1 \end{cases} \tag{VI.2.6}$$

The two radial integrations appearing in eq. (VI.2.3) can be carried out numerically, for example by Gauss-Laguerre quadrature, without any particular troubles arising from the piecewise form of the integrand. It should be noted that the result (VI.2.3) is not restricted to electron charge densities evaluated in the SCF approximation: if more sophisticated atomic wavefunctions are available, they are welcome and one should then use the corresponding values of $\langle\hat{\rho}^{el}(r)\rangle_0$ in eq. (VI.2.3)

The extension to non-spherical systems of the previous approach, however non-trivial even in the case of rather symmetric molecules, is possible. We give here a somewhat detailed review, based on the bipolar expansion method of Buehler and Hirschfelder [1,139,140], recently reconsidered by Meath and coworkers [141].

Let us consider two arbitrary molecules A and B far apart, R being the distance between their centres of mass a and b. If two parallel space-fixed coordinate frames, having their origins in a and b respectively are introduced (see Fig. VI.2), we can take advantage of the following bipolar expansion of $|\underline{r}' - \underline{r}|^{-1}$,

$$\frac{1}{|\underline{r}'-\underline{r}|} = \frac{1}{|\underline{r}_b+\underline{R}-\underline{r}_a|}$$

$$= \sum_{L_a=0}^{\infty} \sum_{L_b=0}^{\infty} \sum_{M=-L_<}^{+L_<} J_{L_aL_b}^{|M|}(r_a,r_b;R)\, Y_{L_a}^{M}(\bar{\underline{\omega}}_a)\, Y_{L_b}^{-M}(\bar{\underline{\omega}}_b) \qquad \text{(VI.2.7)}$$

where $(r_a,\bar{\underline{\omega}}_a) \equiv (r_a,\bar{\theta}_a,\bar{\varphi}_a)$ are the space-fixed spherical components of the position vectors $\underline{r} \equiv \underline{r}_a$, $Y_L^M(\bar{\underline{\omega}})$ is a spherical harmonic of the argument $\bar{\underline{\omega}}$ and $L_<$ the lesser of (L_a,L_b). The coefficients $J_{L_aL_b}^{|M|}$* possess different functional forms in each of the four regions displayed in Fig. VI.3. The limit situation of two non-overlapping charge distribu-

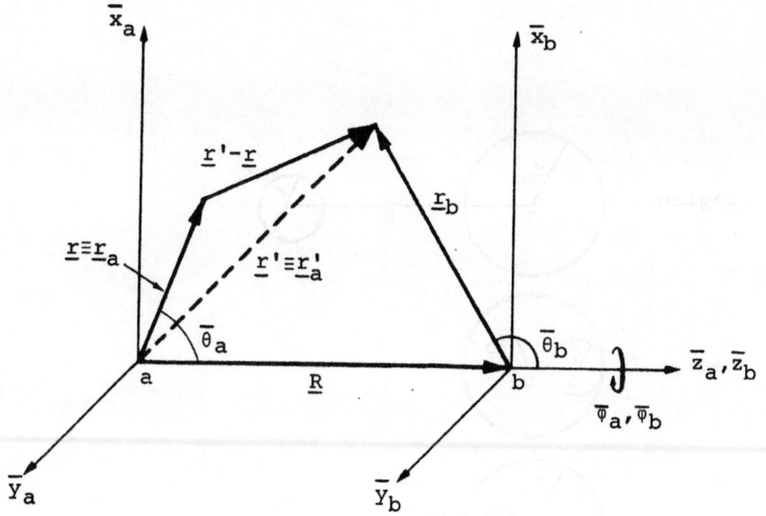

Fig. VI.2. (Space-fixed) coordinate frame used for the bipolar expansion of $|\underline{r}'-\underline{r}|^{-1}$.

---

* The functions $J_{L_aL_b}^{|M|}$ which appear in eq. (VI.2.7) are related to the functions $B_{L_aL_b}^{|M|}$ of Buehler and Hirschfelder by [141]

$$J_{L_aL_b}^{|M|} = 4\pi(-1)^{|M|}\left[\frac{(L_a+|M|)!\,(L_b+|M|)!}{(2L_a+1)(2L_b+1)(L_a-|M|)!\,(L_b-|M|)!}\right]^{\frac{1}{2}} B_{L_aL_b}^{|M|}$$

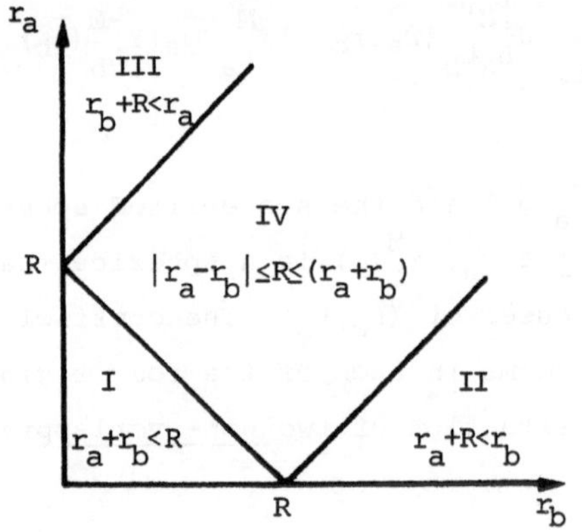

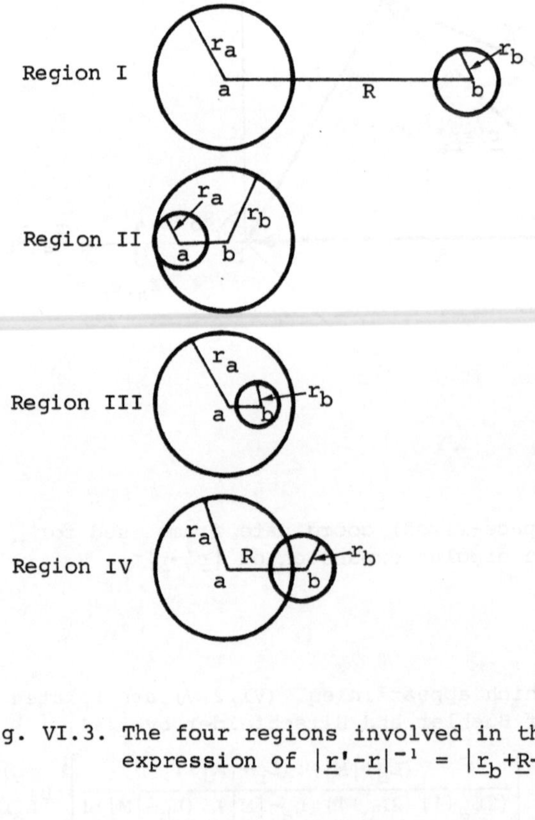

Fig. VI.3. The four regions involved in the bipolar expression of $|\underline{r}'-\underline{r}|^{-1} = |\underline{r}_b+\underline{R}-\underline{r}_a|^{-1}$.

tions corresponds only to the region I, where $r_a+r_b<R$. Analytical expressions for the coefficients $J^{|M|}_{L_aL_b}$ can be obtained for $(L_a,L_b)=0,1,2,3$ from the reference [1].

It is convenient to get an explicit orientation dependence of the Coulomb interaction energy E_{01} between any two molecules. To this end, we introduce <u>body-fixed</u> coordinate axes in each molecule, having a common origin with the corresponding space-fixed axes (Fig. VI.4).

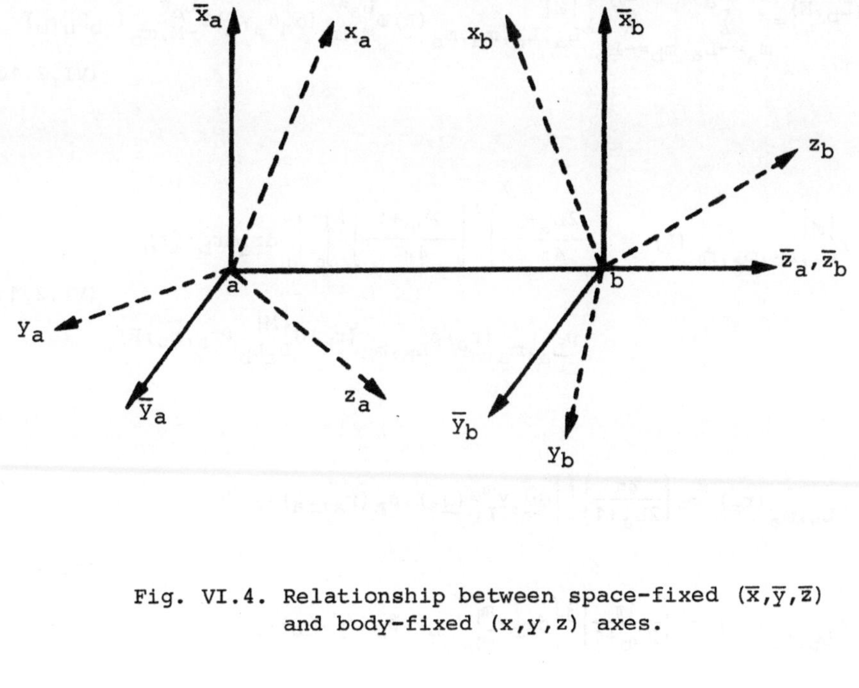

Fig. VI.4. Relationship between space-fixed $(\bar{x},\bar{y},\bar{z})$ and body-fixed (x,y,z) axes.

The relationship between space- and body-fixed spherical harmonics within a molecule, for instance A, results [142,143]

$$Y^M_{L_a}(\bar{\underline{\omega}}_a) = \sum_{m_a=-L_a}^{+L_a} \mathbb{D}^{L_a *}_{M,m_a}(\alpha_a\beta_a\gamma_a)\, Y^{m_a}_{L_a}(\underline{\omega}_a) \qquad \text{(VI.2.8)}$$

where $(\alpha_a\beta_a\gamma_a)$ are the Euler angles which individuate the rotation of the body-fixed axes with respect to the space-fixed ones in the molecule A and $\mathbb{D}^{L_a}(\alpha_a\beta_a\gamma_a)$ is the $(2L_a+1)$-dimensional irreducible representation matrix of the (three-dimensional) rotation group. An analogous expression will hold for $Y^M_{L_b}(\bar{\underline{\omega}}_b)$.

If we substitute eqs. (VI.2.7) and (VI.2.8) into eq. (VI.1.5) we get the following general result

$$E_{01} = \sum_{L_a=0}^{\infty}\sum_{L_b=0}^{\infty} E_{01}^{(L_a,L_b)} = \sum_{L_a=0}^{\infty}\sum_{L_b=0}^{\infty}\sum_{M=-L_<}^{+L_<} E_{01}^{(L_a,L_b,M)} \qquad \text{(VI.2.9)}$$

$$E_{01}^{(L_a,L_b,M)} = \sum_{m_a=-L_a}^{+L_a}\sum_{m_b=-L_b}^{+L_b} \chi^{|M|}_{L_a,L_b,m_a,m_b}(R)\,\mathbb{D}^{L_a*}_{M,m_a}(\alpha_a\beta_a\gamma_a)\,\mathbb{D}^{L_b*}_{-M,m_b}(\alpha_b\beta_b\gamma_b) \qquad \text{(VI.2.10)}$$

where

$$\chi^{|M|}_{L_a,L_b,m_a,m_b}(R) = \left(\frac{2L_a+1}{4\pi}\right)^{\frac{1}{2}}\left(\frac{2L_b+1}{4\pi}\right)^{\frac{1}{2}}\int_0^{\infty}\int_0^{\infty} dr_a dr_b r_a^2 r_b^2 \cdot \rho_{L_a,m_a}(r_a)\,\rho_{L_b,m_b}(r_b)\,J^{|M|}_{L_aL_b}(r_a,r_b;R) \qquad \text{(VI.2.11)}$$

and

$$\rho_{L_a,m_a}(r_a) = \left(\frac{4\pi}{2L_a+1}\right)^{\frac{1}{2}}\int d\underline{\omega}_a Y^{m_a}_{L_a}(\underline{\omega}_a)\langle\hat{\rho}_A(r_a,\underline{\omega}_a)\rangle_0$$

$$\rho_{L_b,m_b}(r_b) = \left(\frac{4\pi}{2L_b+1}\right)^{\frac{1}{2}}\int d\underline{\omega}_b Y^{m_b}_{L_b}(\underline{\omega}_b)\langle\hat{\rho}_B(r_b,\underline{\omega}_b)\rangle_0 \qquad \text{(VI.2.12)}$$

For general molecules, therefore, the Coulomb interaction energy in principle consists of an infinity of contributions, eq. (VI.2.9). The orientation dependence of the various contributions $E_{01}^{(L_a,L_b)}$ or $E_{01}^{(L_a,L_b,M)}$ is explicitly contained in the coefficients $\mathbb{D}^{L}_{M,m'}$, while the intermolecular distance dependence arises from the coefficients $\chi^{|M|}_{L_a,L_b,m_a,m_b}(R)$.

In actual applications, because of generally present symmetries of one or both of the interacting partners, simplifications of the complicate result (V.2.9) become possible. For instance, in the case of interaction between two diatomic molecules, if each body-fixed frame is chosen in such a way the z-axis lies along the internuclear versor $\underline{R}/R$, $\gamma_a = \gamma_b = 0$, $\beta = \theta$ and $\alpha = \varphi$, (θ,φ) being spherical polar angles. For axially symmetric molecules, one has $\rho_{L,m}(r) = \rho_{L,0}(r)\delta_{m,0}$, so that eq. (VI.2.10) simplifies in a noticeable way:

$$E_{01}^{(L_a,L_b,M)} = \chi^{|M|}_{L_a,L_b,0,0}(R)\,C^{M}_{L_a}(\theta_a,\varphi_a)\,C^{-M}_{L_b}(\theta_b,\varphi_b) \qquad \text{(VI.2.13)}$$

We have taken into account that $D^{L}_{M,0}(\alpha,\beta,0) = [4\pi/(2L+1)]^{\frac{1}{2}}\,Y^{M}_{L}(\beta,\alpha)$ [142] and introduced the modified or Racah spherical harmonics [143]

$$C^{M}_{L}(\theta,\varphi) = \left(\frac{4\pi}{2L+1}\right)^{\frac{1}{2}} Y^{M}_{L}(\theta,\varphi)$$

In the case of two interacting diatomic molecules eq. (VI.2.9) becomes

$$E_{01} = \sum_{L_a=0}^{\infty}\ \sum_{L_b=0}^{\infty}\ \sum_{M=-L_<}^{+L_<} \chi^{|M|}_{L_a,L_b,0,0}(R)\,C^{M}_{L_a}(\theta_a,\varphi_a)\,C^{-M}_{L_b}(\theta_b,\varphi_b) \qquad \text{(VI.2.14)}$$

an expression with a manifestly simplified orientation dependence.

Applications of the present formalism to the case of the interaction between two hydrogen molecules have been carried out by Meath and coworkers [141] (see sec. VI.4).

VI.3. The Coulombic Interaction Energy in the Approximation of Neglecting Overlap Effects.

In this section we shall consider the very important case where the two molecular charge distributions are so far apart that overlap effects can be neglected: in terms of the approach developed in the preceding section, this means that the contribution to the interaction energy will totally arise from the region I, where $r_a+r_b<R$, $\forall r_a, r_b$, and simplified expressions for the interaction energy are therefore to

be expected.

It should be clear that in no case two realistic molecular charge distributions are truly non-overlapping, so that we are actually considering an approximation: due to the supposed large distance between the "monomers", we are looking at the two molecules as if their charge distributions rigorously had no penetration.

Very useful expansions for the quantity $|\underline{r}'-\underline{r}|^{-1} = |\underline{r}_b+\underline{R}-\underline{r}_a|^{-1}$ are available in this case [1-4,144,145], which allow the electrostatic interaction energy E_{01} to be expressed in the form of a series in the parameter (1/R). Since each term of such series represents the interaction between an (electric) multipole centered at the first molecule and a multipole centered at the second one, the resulting series is referred to as the <u>multipole expansion</u> of the (electrostatic) interaction energy.

Two standard ways of presenting the multipole expansion make use of <u>Cartesian</u> and <u>irreducible multipole moment tensors</u>, respectively [145]. The first approach has the advantage of an undoubted simplicity of derivation, but provides a redundant final result, because the $\frac{1}{2}(n+1)(n+2)$ Cartesian components of a n-th rank tensor afford representations of the full rotation group (invariance group of $|\underline{r}'-\underline{r}|^{-1}$), which, however, are <u>not</u> irreducible. Irreducible tensors from this point of view offer the most economic and powerful approach, because the $|\underline{r}_b+\underline{R}-\underline{r}_a|^{-1}$ expansion is then expressed in terms of spherical harmonics and Clebsch-Gordan coefficients, with many advantages in the ensuing manipulations.

If two (space-fixed) <u>parallel</u> coordinate frames are introduced in the molecules A and B, their centers being at a and b respectively (the vector $\underline{R}$ points from a to b), the following expansion for $|\underline{r}'-\underline{r}|^{-1}$ can be used [145,146]

$$|\underline{r}'-\underline{r}|^{-1} = |\underline{r}_b+\underline{R}-\underline{r}_a|^{-1} = \sum_{L_a=0}^{\infty} \sum_{L_b=0}^{\infty} R^{-(L_a+L_b+1)} V_{L_aL_b} \qquad \text{(VI.3.1)}$$

where

$$V_{L_aL_b} = (4\pi)^{3/2} (-1)^{L_b} \binom{2L}{2L_a}^{1/2} \left[(2L_a+1)(2L_b+1)(2L+1)\right]^{-1/2}$$

$$\sum_{M=-L}^{+L} (-1)^{M} \cdot Y_L^{-M}(\underline{R}/R)\left[\mathcal{Y}_{L_a}(\underline{r}_a) \boxtimes \mathcal{Y}_{L_b}(\underline{r}_b)\right]_M^L \tag{VI.3.2}$$

In the preceding expression we have put $L = L_a+L_b$ and introduced the regular spherical harmonic

$$\mathcal{Y}_\ell^m(\underline{r}) = r^\ell Y_\ell^m(\hat{r}) \quad , \hat{r} = \underline{r}/r \tag{VI.3.3}$$

The quantity

$$[U_{\ell_1} \boxtimes V_{\ell_2}]_m^\ell \equiv \sum_{m_1 m_2} U_{\ell_1}^{m_1} V_{\ell_2}^{m_2} \langle \ell_1 m_1 ; \ell_2 m_2 | \ell m \rangle \tag{VI.3.4}$$

denotes the irreducible product of the two irreducible tensor sets U_{ℓ_1}, V_{ℓ_2}, $\langle \ell_1 m_1 ; \ell_2 m_2 | \ell m \rangle$ being a Clebsch-Gordan coefficient [145,146]. Therefore $[\mathcal{Y}_{L_a}(\underline{r}_a) \boxtimes \mathcal{Y}_{L_b}(\underline{r}_b)]_M^L$ represents the coupling of a multipole moment of order L_b on B with a multipole moment of order L_a on A, to give a total moment of (maximal) order $L = L_a+L_b$.

If the two reference frames are so chosen that their z-axes lie along $\underline{R}$ (see Fig. VI.4) it is not difficult to show that [1,2,4,144]

$$V_{L_a L_b} = \frac{4\pi(-1)^{L_b}(L_a+L_b)!}{[(2L_a+1)(2L_b+1)]^{1/2}}$$

$$\cdot \sum_{M=-L_<}^{+L_<} \frac{\mathcal{Y}_{L_a}^{M}(\underline{r}_a)\mathcal{Y}_{L_b}^{-M}(\underline{r}_b)}{[(L_a-M)!(L_a+M)!(L_b-M)!(L_b+M)!]^{1/2}} \tag{VI.3.5}$$

$L_<$ being the smaller of L_a and L_b.

In the approximation of neglecting overlap effects, the Coulomb interaction energy between two molecular charge distributions assumes the factorized form

$$E_{01} = \sum_{L_a=0}^{\infty} \sum_{L_b=0}^{\infty} \sum_{M=-L_<}^{+L_<} \frac{4\pi(-1)^{L_b}(L_a+L_b)!}{[(2L_a+1)(2L_b+1)(L_a-M)!(L_a+M)!(L_b-M)!(L_b+M)!]^{1/2}}$$

$$\cdot R^{-(L_a+L_b+1)} \int d\underline{r}_a \mathcal{Y}^{M}_{L_a}(\underline{r}_a)\langle\hat{\rho}_A(\underline{r}_a)\rangle_0 \int d\underline{r}_b \mathcal{Y}^{-M}_{L_b}(\underline{r}_b)\langle\hat{\rho}_B(\underline{r}_b)\rangle_0 \tag{VI.3.6}$$

i.e. a series in inverse powers of (1/R).

The lowest order term of this series, the monopole-monopole one, corresponds to $L_a = L_b = 0$ and is given by $q_A q_B/R$, when we have introduced the (net) electric charge (or monopole moment) $q = \int d\underline{r}\langle\hat{\rho}(\underline{r})\rangle_0$. For any neutral system, $q = 0$, so that contributions to E_{01} of the monopole-multipole type require at least one of the interacting partners to have a non-zero net charge.

The dipole-dipole contributions E^{d-d}_{01} to E_{01} ($L_a=L_b=1$) can very easily be given the following Cartesian form

$$E^{d-d}_{01} = R^{-3}\left[\int d\underline{r}_a x_a\langle\hat{\rho}_A(\underline{r}_a)\rangle_0 \int d\underline{r}_b x_b\langle\hat{\rho}_B(\underline{r}_b)\rangle_0\right.$$

$$\left.+ \int d\underline{r}_a y_a\langle\hat{\rho}_A(\underline{r}_a)\rangle_0 \int d\underline{r}_b y_b\langle\hat{\rho}_B(\underline{r}_b)\rangle_0 - 2\int d\underline{r}_a z_a\langle\hat{\rho}_A(\underline{r}_a)\rangle_0 \int d\underline{r}_b z_b\langle\hat{\rho}_B(\underline{r}_b)\rangle_0\right]$$

If we introduce the permanent dipole moment (vector) $\underline{\mu}$ of a charge distribution, $\underline{\mu} = \int d\underline{r}\,\underline{r}\langle\hat{\rho}(\underline{r})\rangle_0$, we can also write

$$E^{d-d}_{01} = \underline{\mu}_A \cdot (R^2\mathbb{1} - 3\underline{R}\,\underline{R})\cdot\underline{\mu}_B/R^5$$

($\mathbb{1}$ denotes the unit dyadic), the classical result for the mutual potential energy of two electric dipoles $\underline{\mu}_A$ and $\underline{\mu}_B$ [133].

The orientation dependence of the Coulomb interaction energy can explicitly be put in evidence making use of the relation betwwen space and body-fixed spherical harmonics within each molecule, eq. (VI.2.8). The final result assumes the same form as eqs. (VI.2.9), (VI.2.10), the only difference being that now $\chi^{|M|}_{L_aL_bm_am_b}(R)$ has the following simplified expression

$$\tilde{\chi}^{|M|}_{L_aL_bm_am_b}(R) = \frac{(-1)^{L_b}(L_a+L_b)!\,R^{-(L_a+L_b+1)}}{[(L_a-|M|)!(L_a+|M|)!(L_b-|M|)!(L_b+|M|)!]^{1/2}}$$

$$\cdot \int_0^\infty dr_a\, r_a^{L_a+2}\ \rho_{L_a m_a}(r_a) \int_0^\infty dr_b\, r_b^{L_b+2}\ \rho_{L_b m_b}(r_b) \qquad \text{(VI.3.7)}$$

the radial density charge component $\rho_{Lm}(r)$ being defined by eq. (VI.2.12). It is therefore apparent how the knowledge of the permanent 2^L-pole moments Q_L^m of the involved charge distributions

$$Q_L^m = \int_0^\infty dr\ r^{L+2}\ \rho_{Lm}(r) \qquad \text{(VI.3.8)}$$

allows to avaluate $\tilde{\chi}^{|M|}_{L_aL_bm_am_b}(R)$ and consequently $E_{01}(R)$, for any relative orientation of the two molecules*.

Simplifications of the general result, eq. (VI.3.7), will clerly arise when symmetry properties can be invoked. For example, for axially symmetric molecules, $Q_L^m = Q_L^0\delta_{m0}$, as a consequence of the fact that $\rho_{Lm}(r) = \rho_{L0}(r)\delta_{m0}$; for spherical atoms, $\rho_{Lm}(r) = \rho_{00}(r)\delta_{L0}\delta_{m0}$ and only Q_0^0 will be eventually non-vanishing. We learn therefore that the Coulombic interaction energy between neutral spherical atoms arises completely from charge-overlap effects.

VI.4. An Assessment of the Importance of Charge Overlap Effects.

Since the neglect of overlap effects between the interacting subsystems leads to noticeable computational simplifications, it is important to gain some feeling about the limit of applicability and reliability of the approximation to E_{01} based on the multipole expansion (see preceding section). Although no fully general result is available that can easily be applied to any situation, we are today in a position to analyze a number of specific examples, which provide us with some pieces of experience. Meath *and coll.* have been particularly active to this regard [92,141,147-150] and just from their results we are about to draw some conclusions.

---

*It is immediately seen that $Q_0^0 = q$, the net electric charge previously introduced. Analogously, for the electric dipole moment $\underline{\mu}$ one finds $\mu_x = -2^{-1/2}(Q_1^1 - Q_1^{-1})$, $\mu_y = +2^{-1/2}\,i(Q_1^1 + Q_1^{-1})$, $\mu_z = Q_1^0$.

The interaction $H^+....NH_3$ is a first good example, from which we may hope to learn something, including the anisotropic properties of more general cases [149]. If we neglect charge overlap effects, from eqs. (VI.1.5), (VI.3.1), (VI.3.2) and (VI.3.8) we get in a straightforward way

$$E_{01} = (4\pi)^{1/2} \sum_{L_a} \sum_{L_b} (-1)^{L_b} \binom{2L}{2L_a}^{1/2} [(2L+1)]^{-1/2} R^{-(L_a+L_b+1)}$$

$$\cdot \sum_{M=-L}^{+L} (-1)^M Y_L^{-M}(\underline{R}/R) \sum_{M_a M_b} \langle L_b M_b; L_a M_a | LM\rangle \bar{Q}_{L_b}^{M_b} \bar{Q}_{L_a}^{M_a}$$

where $\bar{Q}_{L_c}^{M_c}$ is the permanent 2^L-pole moment of the subsystem C referred to space-fixed axes and $\underline{R} \equiv (R, \underline{R}/R)$ the position vector of the proton H^+ measured from the center of mass of the molecule NH_3 (where we have chosen the origin of our coordinate frame). For a proton, $\bar{Q}_L^M = \delta_{L0}\delta_{M0}$, so that after some simple manipulations,

$$E_{01} = \sum_{L_a=1}^{\infty} \sum_{M=-L_a}^{+L_a} \left(\frac{4\pi}{2L_a+1}\right)^{1/2} R^{-(L_a+1)} Y_{L_a}^M(\underline{R}/R) \bar{Q}_{L_a}^{M*} \qquad \text{(VI.4.1)}$$

In order to make explicit the orientation dependence of the interaction energy, we express the space-fixed multipole moments $\bar{Q}_L^M$ of the NH_3 molecule in terms of the multipole moments Q_L^m defined in the body-fixed frame,

$$\bar{Q}_{L_a}^M = \sum_m D_{M,m}^{L_a *}(\alpha_a, \beta_a, \gamma_a) Q_{L_a}^m$$

(see eqs. (VI.2.8) and (VI.3.3)). If the body-fixed coordinate frame in NH_3 is chosen with the Z-axis along the molecular three-fold symmetry axis, oriented from the center of mass to the N atom, and the X-axis in a vertical plane containing a NH bond, then

$$Q_1^m = \mu\delta_{m0} \quad , \qquad Q_2^m = Q\delta_{m0} \quad , \qquad \text{(VI.4.2)}$$

if we limit ourselves to retain the first two molecular multipole moments (that seems realistic enough, in view of our rather uncomplete knowledge of higher moments for most molecules. In this way,

$$E_{01} = \left(\frac{4\pi}{3}\right)^{1/2} \frac{\mu}{R^2} \sum_M Y_1^M(\underline{R}/R)\, D^1_{M,0}(\alpha_a,\beta_a,\gamma_a)$$

$$+ \left(\frac{4\pi}{5}\right)^{1/2} \frac{Q}{R^3} \sum_M Y_2^M(\underline{R}/R)\, D^2_{M,0}(\alpha_a,\beta_a,\gamma_a) + O(R^{-4})$$

A convenient final simplification is attained by choosing space- and body-fixed frames in the molecule aligned, so that $D^L_{M,m} \to \delta_{Mm}$; then,

$$E_{01} = \frac{\mu}{R^2} \cos\theta + \frac{Q}{2R^2} (3\cos^2\theta - 1) + O(R^{-4}) \qquad \text{(VI.4.3)}$$

where (θ,Φ) are the polar angles which individuate the unit vector $\underline{R}/R$ and $O(R^{-4})$ denotes that we are neglecting terms depending at lowest order on the octopole moment of NH_3. (We have chosen to give a rather detailed derivation of this result as an example of manipulation of the general multipole expansion, eq. (VI.3.6)).

As a simple (and interesting) application of eq. (VI.4.3) let us assume that the collision $NH_3 ... H^+$ occurs with the proton moving along a straight line trajectory (see Fig. (VI.5)), i.e.

$$\underline{R} = \underline{b} + \lambda\underline{a}$$

where $\underline{b}$ is a vector trough the origin, perpendicular to the collision path ($b \equiv |\underline{b}|$ is nothing but the classical <u>impact parameter</u> for the collision), while $\lambda\underline{a}$ is the displacement vector along the trajectory ($\underline{a}$ is a unit vector). Taking into account that $\underline{R} \equiv (R\text{sen}\theta\cos\Phi, R\text{sen}\theta\,\text{sen}\Phi, R\cos\theta)$ and $\underline{a} \equiv (\text{sen}\theta\cos\phi, \text{sen}\theta\,\text{sen}\phi, \cos\theta)$, from $\lambda = [R^2-b^2]^{1/2} = R[1 - (b/R)^2]^{1/2}$ one finds easily the following alternative expression for E_{01}, eq. (VI.4.3), in the case $R > b$ (so that λ can be expanded in powers of R^{-1})

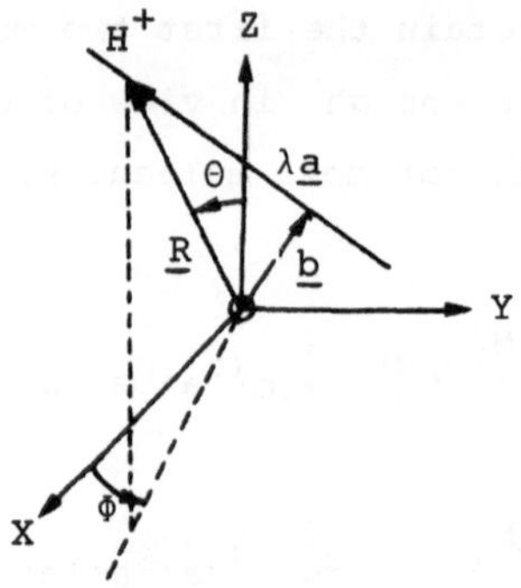

Fig. VI.5. Linear collision trajectory for the interaction $NH_3 \ldots H^+$ and involved parameters.

$$E_{01} = \frac{\mu}{R^2}\cos\theta + \frac{1}{R^3}\left[\mu b_z + \frac{Q}{2}(3\cos^2\theta - 1)\right] + \frac{1}{R^4}\left(3Qb_z - \frac{\mu b^2}{2}\right)\cos\theta + O(R^{-5}) \qquad \text{(VI.4.4)}$$

(the R^{-4} coefficient is not exact, because terms involving the octopole moment Q_3^m actually contribute at the same order; the missing part $+\frac{1}{2}R^{-4}[Q_3^0(5\cos^3\theta-3\cos\theta)-5^{1/2}\, Q_3^3 \mathrm{sen}^3\theta\cos 3\phi]$, is clearly ϕ-dependent).

In Table VI.1 we report some results for the interaction $NH_3 \ldots H^+$, as the proton moves along some typical trajectories. E_{01}^{exp} and $E_{01}^{non-exp}$ represent respectively the 1-st order Coulomb energy in the expanded form, eq. (VI.4.4), and the non-expanded result, i.e. that obtained without making recourse to the multipole expansion, eq. (VI.3.1). The values for both $E_{01}^{non-exp}$ and E_{01}^{exp} have been calculated from the energetically accurate Clementi's SCF wavefunction [151] for the ground state of NH_3 in its equilibrium geometry. Although the evaluated dipole moment μ_{SCF} = -0.90363 a.u. and quadrupole moment Q_{SCF} = -2.08405

Table VI.1. Coulombic contribution to the 1-st order interaction energy for $NH_3...H^+$ as a function of the distance R: comparison of the non-expanded result with the multipole expansion through charge-quadrupole terms.

	$\underline{b}=0$, $\theta=\theta=0$		$\underline{b}=0$, $\theta=\theta=\pi$	
R(a.u.)	$-E_{01}^{exp}$	$-E_{01}^{non-exp}$	E_{01}^{exp}	$E_{01}^{non-exp}$
2	0.4864 (0)	0.1484 (0)	-0.3460 (-1)	0.5943 (-1)
4	0.8904 (-1)	0.8000 (-1)	0.2391 (-1)	0.1863 (-1)
6	0.3475 (-1)	0.3320 (-1)	0.1545 (-1)	0.1393 (-1)
8	0.1819 (-1)	0.1768 (-1)	0.1005 (-1)	0.9533 (-2)
10	0.1112 (-1)	0.1091 (-1)	0.6952 (-2)	0.6735 (-2)
20	0.2520 (-2)	0.2506 (-2)	0.1999 (-2)	0.1985 (-2)

	$\underline{b}=0$, $\theta=\theta=\cos^{-1}\left(\frac{3}{5}\right)^{1/2}$, $\phi=\Phi=\frac{\pi}{6}$			$b_x=b_y=0$, $\lambda\neq R$, $\theta\neq\theta=\pi/2$; $b_z=Q/2$, $\phi=\Phi=\pi$	
R(a.u.)	$-E_{01}^{exp}$	$-E_{01}^{non-exp}$	λ(a.u.)	$-E_{01}^{exp}$	$-E_{01}^{non-exp}$
2	0.2752 (0)	0.7307 (-1)	2	0.6339 (-1)	0.3863 (-1)
4	0.5677 (-1)	0.5463 (-1)	4	0.3325 (-2)	0.1106 (-1)
6	0.2330 (-1)	0.2321 (-1)	6	0.4882 (-3)	0.2583 (-2)
8	0.1257 (-1)	0.1255 (-1)	8	0.1205 (-3)	0.8420 (-3)
10	0.7833 (-2)	0.7829 (-2)	10	0.4022 (-4)	0.3451 (-3)
20	0.1854 (-2)	0.1854 (-2)	20	0.1288 (-5)	0.2073 (-4)

a.u. do not compare too favourably with the corresponding experimental values (μ_{expt} = -0.58 a.u., Q_{expt} = -0.744 a.u.), the content of Table V.1 is still meaningful if one is satisfied with comparisons between $E_{01}^{non-exp}$ and E_{01}^{exp}.

The agreement between the two quantities under investigation becomes bad for any trajectory, as the distance R between proton and NH_3 molecule decreases more and more, a consequence of the <u>divergent</u> nature of the multipole expansion. The marked anisotropic behaviour of the expansion is not unexpected because the various charge-multipole contributions add to each other with different signs according to the trajectory considered. Along the trajectories ($\underline{b}=0$, $\theta=\theta=0$ or π) the fairly good representation of $E_{01}^{\text{non-exp}}$ by E_{01}^{exp} at not too small distances is related to the fact that E_{01}^{exp} is correct through terms in R^{-3} (see eq. VI.4.4)). The trajectory ($\underline{b}=0$, $\theta=\cos^{-1}\left(\frac{3}{5}\right)^{1/2}$, $\phi=\pi/6$) is a particularly lucky one, because in this case E_{01}^{exp} is correct through terms in R^{-4} (the octopole contributions vanish). Finally, the last trajectory considered ($b_x=b_y=0$, $b_z=Q/2\mu$, $\theta=\pi/2$, $\phi=0$) offers an example of bad agreement between E_{01}^{exp} and $E_{01}^{\text{non-exp}}$, even at large distances. Because of the particular choice of the parameters, all of the terms retained in eq. (VI.4.4) do not contribute to E_{01}^{exp} and therefore $E_{01}^{\text{exp}} = O(R^{-5})$; but the charge-octopole contributions neglected vary as $O(R^{-4})$, so that the error is larger than E_{01}^{exp} itself, with consequent unreliable results for E_{01}.

The second example we propose to examine in some detail is the interaction $H_2 \ldots H_2$ [141]. The 1-st order Coulombic energy is conveniently expressed by eq. (VI.2.14), the R-dependence being rigorously contained in the coefficients $\chi^{|M|}_{L_a,L_b,0,0}(R)$, eq. (VI.2.11), or approximately in the coefficients $\tilde{\chi}^{|M|}_{L_a,L_b,0,0}(R)$, eq. (VI.3.7), if the multipole expansion is used.

In Table VI.2 we report a number of 1-st order coefficients $\chi^{|M|}_{L_a,L_b}(R) \equiv \chi^{|M|}_{L_a,L_b,0,0}(R)$ and $\tilde{\chi}^{|M|}_{L_a,L_b}(R)$ as obtained by Meath and coll. [141] by using an extremely accurate SCF wavefunction calculated by Davidson and Jones [152] (essentially the Hartree-Fock wavefunction for H_2). It should be noted that i) $\chi^{|M|}_{L_a,L_b}(R) = 0$ if either L_a or L_b is odd, ii) $\chi^{0}_{0,2}(R) = \chi^{0}_{2,0}(R)$ and iii) $\tilde{\chi}^{|M|}_{L_a,L_b}(R) = 0$ if either L_a or L_b is zero.

The first non-zero contribution to E_{01} for the interaction $H_2 \ldots H_2$ as the multipole expansion is used corresponds to the quadrupole-quadrupole energy, which varies as R^{-5}. From the inspection of the Table it is apparent how entirely reliable results may be obtained from the expanded form of E_{01} only at not too small R values ($R \gtrsim 4$ a.u.): at smaller distances, the charge overlap role becomes important, even for a correct description of the anisotropy features. The contributions to

Table VI.2. Calculated $\chi^{|M|}_{L_a,L_b}(R)$ and $\tilde{\chi}^{|M|}_{L_a,L_b}(R)$ coefficients for the interaction $H_2 \ldots H_2$.

R(a.u.)	2.0	3.0	4.0	5.0	6.0	7.0	8.0
$-\chi^0_{0,0}(R)$	3.636(-2)	1.668(-2)	3.369(-3)	5.597(-4)	8.480(-5)	1.215(-5)	1.674(-5)
$-\chi^0_{0,2}(R)$	-6.689(-2)	-1.285(-3)	3.831(-4)	9.740(-5)	1.694(-5)	2.575(-6)	3.642(-7)
$-\chi^0_{2,2}(R)$	1.350(-1)	9.696(-3)	1.545(-3)	4.596(-4)	1.854(-4)	8.662(-5)	4.463(-5)
$-\tilde{\chi}^0_{2,2}(R)$	0.458(-1)	6.031(-3)	1.431(-3)	4.688(-4)	1.884(-4)	8.714(-5)	4.471(-5)
$-\chi^1_{2,2}(R)$	7.051(-2)	5.098(-3)	9.623(-4)	3.079(-4)	1.247(-4)	5.799(-5)	2.979(-5)
$-\tilde{\chi}^1_{2,2}(R)$	3.053(-2)	4.021(-3)	9.540(-4)	3.125(-4)	1.256(-4)	5.809(-5)	2.981(-5)
$-\chi^2_{2,2}(R)$	1.312(-2)	1.093(-3)	2.367(-4)	7.757(-5)	3.133(-5)	1.452(-5)	7.452(-6)
$-\tilde{\chi}^2_{2,2}(R)$	0.763(-2)	1.005(-3)	2.385(-4)	7.813(-5)	3.140(-5)	1.452(-5)	7.452(-6)

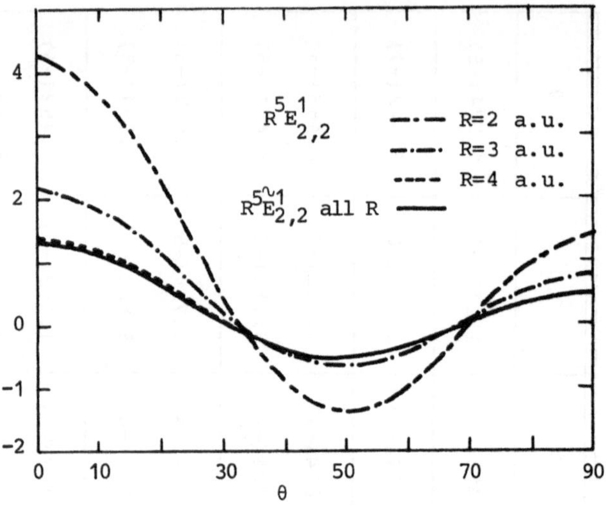

Fig. VI.6. Quadrupole-quadrupole energy for the interaction $H_2...H_2$: a comparison between non-expanded and expanded form results. ($\theta=\theta_a=\theta_b$, $\phi_a=\phi_b=0$).

the interaction energy which drop to zero in an exponential way, so that they are missing in the multipole expansion, are seen from Table VI.1 to be not entirely negligible.

A rather vivid picture of the anisotropic behaviour as described by $\chi^{|M|}_{L_a,L_b}(R)$ with respect to $\tilde{\chi}^{|M|}_{L_a,L_b}(R)$ is offered in Fig VI.6, which shows the global quadrupole-quadrupole energy for the interaction H_2.. ..H_2 at some intermolecular distances R as a function of the θ angle for the case $\theta_a=\theta_b=\theta$, $\phi_a=\phi_b=0$ (see eq. (VI.2.14)).

As a comment for more complex interactions than $H_2...H_2$, the charge overlap effects are expected to become apparent at much larger values of R relative to the Van der Waals minimum than for the interaction $H_2...H_2$ here considered, as a consequence of the fact that other molecules than H_2 will generally be characterized by less concentrated and more strongly orientation dependent charge distributions.

Some remarks about the dependence of the results on the quality of the wavefunctions employed will be postponed up to sec.VI.6.

VI.5. The Exchange Contribution to the 1-st Order Interaction Energy: an Approximate Evaluation.

After discussing the Coulombic component of the 1-st order interaction energy, we now pass to consider the other basic ingredient, the exchange contribution (see eq. (V.3.8)),

$$E_{01,ex} = \frac{\langle\varphi_0|\hat{V}\hat{\mathcal{O}}^{AB}|\varphi_0\rangle - E_{01}\langle\varphi_o|\hat{\mathcal{O}}^{AB}|\varphi_0\rangle}{1 + \langle\varphi_o|\hat{\mathcal{O}}^{AB}|\varphi_0\rangle} \tag{VI.5.1}$$

restricting ourselves to the case where the ground states of both of the "monomers" A and B are approximately described by single Slater determinants, $|\phi_0^A\rangle$ and $|\phi_0^B\rangle$ respectively. Although more general treatments, valid for arbitrary states $|\varphi_0^A\rangle$ and $|\varphi_0^B\rangle$ are available [153], they will be left out from the present, rather elementary presentation. We shall moreover limit ourselves to an approximate evaluation of $E_{01,ex}$ by retaining in the sum of inter-partner permutations $\hat{\mathcal{O}}^{AB}$ only $\hat{P}^{AB}_{(1)}$, i.e. the permutations which exchange one electron label between the two "monomers" [6,154,155]; the results through $\hat{P}^{AB}_{(2)}$, somewhat more involved, can be found elsewhere [103,156].

Following the detailed analysis by Claverie [6], we note that the two approximate eigenstates $|\phi_0^A\rangle$ and $|\phi_0^B\rangle$ can be expressed in the form

$$|\phi_0^A\rangle = (N^A!)^{\frac{1}{2}}\hat{A}^A[|\chi_1^A(1)\rangle\ldots|\chi_\alpha^A(\alpha)\rangle\ldots|\chi_{N^A}^A(N^A)\rangle] \equiv (N^A!)^{\frac{1}{2}}\hat{A}^A|w_0^A\rangle,$$

$$|\phi_0^B\rangle = (N^B!)^{\frac{1}{2}}\hat{A}^B[|\chi_1^B(N^A+1)\rangle\ldots|\chi_\beta^B(N^A+\beta)\rangle\ldots|\chi_{N^B}^B(N^A+N^B)\rangle] \equiv (N^B!)^{\frac{1}{2}}\hat{A}^B|w_0^B\rangle$$

where $|\chi_\gamma^C(\gamma)\rangle$ denotes the (occupied) γ-th spinorbital of the "monomer" C and $|w_0^C\rangle$ the Hartree product of all the occupied spinorbitals relative to C. Now

$$\hat{P}^{AB}_{(1)} = \sum_{\alpha=1}^{N^A} \sum_{\beta=N^A+1}^{N^A+N^B} \hat{P}_{\alpha\beta} = \sum_{\alpha=1}^{N^A} \sum_{\beta=1}^{N^B} \hat{P}_{\alpha,N^A+\beta}$$

is symmetric with respect to the subsets $\{\alpha=1,\ldots N^A\}$ and $\{\beta=N^A+1,\ldots N^A+N^B\}$ and commutes with any intra-monomer permutation [see also sec (V.3)], so that

$$\hat{P}^{AB}_{(1)} |\phi^A_o \phi^B_o> = (N^A!)^{\frac{1}{2}} (N^B!)^{\frac{1}{2}} \hat{A}^A \hat{A}^B \hat{P}^{AB}_{(1)} |w^A_o w^B_o> \qquad \text{(VI.5.2)}$$

We can now verify in a straightforward way that

$$\hat{P}_{\alpha,N^A+\beta} |w^A_o w^B_o> = \hat{P}_{\alpha,N^A+\beta} \{|\chi^A_1(1)>..|\chi^A_\alpha(\alpha)>..|\chi^B_1(N^A+1)>..|\chi^B_\beta(N^A+\beta)>..\}$$

$$= (|\chi^A_1(1)>..|\chi^B_\beta(\alpha)>..|\chi^A_{N^A}(N^A)>) \times (|\chi^B_1(N^A+1)>..|\chi^A_\alpha(N^A+\beta)>$$

$$..|\chi^B_{N^B}(N^A+N^B)>) = |w^A_o\left(\frac{\beta_B}{\alpha_A}\right)>|w^B_o\left(\frac{\alpha_A}{\beta_B}\right)> \qquad \text{(VI.5.3)}$$

where the last notation means that in the Hartree product $|w^A_o>$ the α-th spinorbital has been replaced by the β-th one belonging to B and in the Hartree product $|w^B_o>$ the β-th spinorbital has been replaced by the α-th one belonging to A. Thus, from eq. (VI.5.2),

$$\hat{P}^{AB}_{(1)} |\phi^A_o \phi^B_o> = \sum_{\alpha=1}^{N^A} \sum_{\beta=1}^{N^B} |\phi^{\beta_B}_{\alpha_A}>|\phi^{\alpha_A}_{\beta_B}> \qquad \text{(VI.5.4)}$$

where $|\phi^{\beta_B}_{\alpha_A}>$ is a single excited configuration (Slater determinant), obtained from $|\phi^A_o>$ by replacing the occupied α_A-th spinorbital by the β_B-th spinorbital occupied in $|\phi^B_o>$.

The single excited Slater determinants $|\phi^{\beta_B}_{\alpha_A}>$, $|\phi^{\alpha_A}_{\beta_B}>$ introduced in this way involve spinorbitals not orthogonal to each other (in fact, $<\chi^A_\alpha|\chi^A_{\alpha'}> = \delta_{\alpha\alpha'}$, $<\chi^B_\beta|\chi^B_{\beta'}> = \delta_{\beta\beta'}$, but $<\chi^A_\alpha|\chi^B_\beta> = S^{AB}_{\alpha\beta} \neq \delta_{\alpha\beta}$) while for the next manipulations it would be convenient to dispose of orthogonal spinorbitals. To this end, let us decompose the row $|\boldsymbol{\chi}^A\rangle \equiv \{|\chi_1^A\rangle,\ldots,|\chi_{N^A}^A\rangle\}$ of spinorbitals belonging to the "monomer" A into its component in the subspace spanned by $|\boldsymbol{\chi}^B\rangle$ and its component in the orthogonal subspace, with an entirely analogous procedure for the row $|\boldsymbol{\chi}^B\rangle$. If $\mathbb{S}^{AB} = \langle\boldsymbol{\chi}^A|\boldsymbol{\chi}^B\rangle$ denotes the $N^A \times N^B$ overlap matrix between the two sets of spinorbitals $|\boldsymbol{\chi}^A\rangle$ and $|\boldsymbol{\chi}^B\rangle$, then

$$|\boldsymbol{\chi}^A\rangle' = |\boldsymbol{\chi}^A\rangle - |\boldsymbol{\chi}^B\rangle\,\mathbb{S}^{BA} \qquad \text{(VI.5.5)}$$

represents the component of $|\boldsymbol{\chi}^A\rangle$ in the orthogonal subspace to that spanned by $|\boldsymbol{\chi}^B\rangle$. In terms of the $N^B \times N^A$ overlap matrix $\mathbb{S}^{BA} = \langle\boldsymbol{\chi}^B|\boldsymbol{\chi}^A\rangle$ we may write the following expression for $|\boldsymbol{\chi}^B\rangle'$, the component of $|\boldsymbol{\chi}^B\rangle$ in the orthogonal subspace to that spanned by $|\boldsymbol{\chi}^A\rangle$

$$|\boldsymbol{\chi}^B\rangle' = |\boldsymbol{\chi}^B\rangle - |\boldsymbol{\chi}^A\rangle\,\mathbb{S}^{AB} \qquad \text{(VI.5.5')}$$

If we now express in the Slater determinant $|\phi_{\alpha_A}^{\beta_B}\rangle$ the spin-orbital $|\chi_\beta^B\rangle$ in the form

$$|\chi_\beta^B\rangle = |\chi_\beta^B\rangle' + \sum_{\alpha'=1}^{N^A} |\chi_{\alpha'}^A\rangle\, S_{\alpha'\beta}^{AB}$$

it follows

$$|\phi_{\alpha_A}^{\beta_B}\rangle = |\phi_{\alpha_A}^{\beta_B}\rangle' + \sum_{\alpha'=1}^{N^A} S_{\alpha'\beta}^{AB}\,|\phi_{\alpha_A}^{\alpha'_A}\rangle$$

where $|\phi_{\alpha_A}^{\beta_B}\rangle'$ is the Slater determinant obtained from $|\phi_0^A\rangle$ by the substitution $|\chi_\beta^B\rangle' \rightarrow |\chi_\alpha^A\rangle$. Now $|\phi_{\alpha_A}^{\alpha'_A}$ i.e. the determinant where $|\chi_\alpha^A\rangle$ has been replaced by $|\chi_{\alpha'}^A\rangle$, vanishes identically unless $\alpha' = \alpha$; furthermore $|\phi_{\alpha_A}^{\alpha_A}\rangle \equiv |\phi_0^A\rangle$, so that

$$|\phi_{\alpha_A}^{\beta_B}\rangle = |\phi_{\alpha_A}^{\beta_B}\rangle' + S_{\alpha\beta}^{AB}\,|\phi_0^A\rangle \qquad \text{(VI.5.6)}$$

Analogously

$$|\phi^{\alpha_A}_{\beta_B}\rangle = |\phi^{\alpha_A}_{\beta_B}\rangle' + S^{AB}_{\alpha\beta}|\phi^B_o\rangle \qquad \text{(VI.5.6')}$$

Coming back to eq. (VI.5.4), we find from eqs. (VI.5.6), (VI.5.6')

$$\hat{P}^{AB}_{(1)}|\phi^A_0\phi^B_o\rangle = \sum_{\alpha=1}^{N^A}\sum_{\beta=1}^{N^B}\{|\phi^{\beta_B}_{\alpha_A}\rangle'|\phi^{\alpha_A}_{\beta_B}\rangle' + S^{AB}_{\alpha\beta}\Big[|\phi^A_o\rangle|\phi^{\alpha_A}_{\beta_B}\rangle'$$

$$+ |\phi^{\beta_B}_{\alpha_A}\rangle'|\phi^B_o\rangle\Big] + (S^{AB}_{\alpha\beta})^2|\phi^A_0\rangle|\phi^B_o\rangle\} \qquad \text{(VI.5.7)}$$

which allows the matrix elements $\langle\phi^A_0\phi^B_o|\hat{V}\hat{P}^{AB}_{(1)}|\phi^A_o\phi^B_o\rangle$, $\langle\phi^A_o\phi^B_o|\hat{P}^{AB}_{(1)}|\phi^A_0\phi^B_o\rangle$ to be readily evaluated. The expression of the matrix element $\langle\phi^A_0\phi^B_o|$ $\hat{P}^{AB}_{(1)}|\phi^A_0\phi^B_o\rangle$ is particularly simple, for the explicit orthogonality of the spinorbitals involved reduces drastically the number of possible contributions:

$$\langle\phi^A_0\phi^B_o|\hat{P}^{AB}_{(1)}|\phi^A_0\phi^B_o\rangle = \sum_{\alpha=1}^{N^A}\sum_{\beta=1}^{N^B}(S^{AB}_{\alpha\beta})^2 \qquad \text{(VI.5.8)}$$

For the matrix element $\langle\phi^A_0\phi^B_o|\hat{V}\hat{P}^{AB}_{(1)}|\phi^A_0\phi^B_o\rangle$ we get in a similar way

$$\langle\phi^A_0\phi^B_o|\hat{V}\hat{P}^{AB}_{(1)}|\phi^A_0\phi^B_o\rangle = \sum_{\alpha=1}^{N^A}\sum_{\beta=1}^{N^B}\{\langle\phi^A_0\phi^B_o|\hat{V}^{AB}_{ee}$$

$$+ S^{AB}_{\alpha\beta}[\langle\phi^B_o|\hat{V}^B_{en}+\langle\phi^A_o|\hat{V}^{AB}_{ee}|\phi^A_o\rangle\;|\phi^{\alpha_A}_{\beta_B}\rangle' \;+ \langle\phi^A_o|\hat{V}^A_{en}+\langle\phi^B_o|\hat{V}^{AB}_{ee}|\phi^B_o\rangle\;|\phi^{\beta_B}_{\alpha_A}\rangle'$$

$$+ (S^{AB}_{\alpha\beta})^2[\langle\phi^A_o|\hat{V}^A_{en}|\phi^A_o\rangle + \langle\phi^B_o|\hat{V}^B_{en}|\phi^B_o\rangle + \langle\phi^A_o\phi^B_o|\hat{V}^{AB}_{ee}|\phi^A_0\phi^B_o\rangle]\} \qquad \text{(VI.5.9)}$$

where we have used eq. (V.3.12), $\hat{V} = \hat{V}^A_{en}+\hat{V}^B_{en}+\hat{V}^{AB}_{ee}$, for the interaction potential energy operator $\hat{V}$.

$\hat{V}^{B}_{en}+\langle\phi^{A}_{o}|\hat{V}^{AB}_{ee}|\phi^{A}_{o}\rangle$ is clearly the overall potential energy operator for the electrons of the subsystem B due to the presence of the unperturbed subsystem A, with an analogous meaning for the quantity $\hat{V}^{A}_{en}+\langle\phi^{B}_{o}|\hat{V}^{AB}_{ee}|\phi^{B}_{o}\rangle$ (see sec.V.3).

In the approximation that $\hat{\mathcal{P}}^{AB} = -\hat{P}^{AB}_{(1)}$, the exchange contribution to the 1-st order interaction energy can be written

$$E^{(1)}_{01,ex} = \frac{X^{AB}_{(1)}}{1 + S^{AB}_{(1)}} \qquad \text{(VI.5.10)}$$

where

$$S^{AB}_{(1)} = - \sum_{\alpha=1}^{N^A} \sum_{\beta=1}^{N^B} (S^{AB}_{\alpha\beta})^2 \qquad \text{(VI.5.8')}$$

The numerator $X^{AB}_{(1)}$ after a few manipulations of eq. (VI.5.9) can be expressed as

$$X^{AB}_{(1)} = - \sum_{\alpha=1}^{N^A} \sum_{\beta=1}^{N^B} \{\langle\phi^{A}_{0}\phi^{B}_{o}|\hat{V}^{AB}_{ee}|(|\phi^{\beta_B}_{\alpha_A}\rangle'|\phi^{\alpha_A}_{\beta_B}\rangle') + S^{AB}_{\alpha\beta}[\langle\phi^{A}_{o}|(\hat{V}^{A}_{en} \qquad \text{(VI.5.11)}$$

$$+ \langle\phi^{B}_{o}|\hat{V}^{AB}_{ee}|\phi^{B}_{o}\rangle)|\phi^{\beta_B}_{\alpha_A}\rangle' + \langle\phi^{B}_{o}|(\hat{V}^{B}_{en} + \langle\phi^{A}_{o}|\hat{V}^{AB}_{ee}|\phi^{A}_{o}\rangle)|\phi^{\alpha_A}_{\beta_B}\rangle']\}$$

Although it is a rather simple task to put eq. (VI.5.11) in a form involving spinorbitals [6,103,153], we shall not transform it any further.

Some insight into the role of multiple exchange effects between two interacting subsystems can be gained through the inspection of results obtained for some typical interactions. In Table VI.3, the 1-st order energy for the interaction between two near atoms in their ground states is analyzed, so as to allow some conclusions about the convergence of a procedure where the multiple exchanges are only partially taken into account [156]. The Coulombic portion E_{01} of the 1-st order

Table VI.3. Ne-Ne 1-st order interaction energy (a.u.): role of multiple exchange effects.

R(a.u.)	E_{01} ($\times 10^2$)	$E^{(1)}_{01,ex}$ ($\times 10^2$)	$E^{(2)}_{01,ex}$ ($\times 10^2$)	$E_{01,ex}$ ($\times 10^2$)
2.5	-5.043	36.324	-2.855	33.493
3.0	-0.675	7.227	-0.104	7.123
3.5	-0.067	1.331	-0.004	1.327
4.0	-0.014	0.221	0.000	0.221
5.0	-0.000	0.007	0.000	0.007

energy is attractive in the range 2.5<R<5.0 a.u., but is dominated by the exchange energy $E_{01,ex}$ (values in the last column), so that the 1-st order interaction becomes repulsive over the whole range. $E^{(n)}_{01,ex}$ represents the contribution to $E_{01,ex}$ from $P^{AB}_{(n)}$, i.e. from n-exchange interactions between the two Ne atoms: the goodness of the approximation $E_{01,ex} \sim E^{(1)}_{01,ex} + E^{(2)}_{01,ex}$ over the whole range of distances and the adequacy of $E_{01,ex} \sim E^{(1)}_{01,ex}$ in a substantial part of the same range are immediately recognized.

The values collected in Table VI.3 should be considered rather cautiously for truly quantitative assessments. They correspond, in fact,to a calculation by a SCF-wavefunction built-up in terms of a minimal basis set, which is known to afford in general only a modest approximation to the true H.F. wavefunction. The behaviour of this kind of basis set proves to be particularly unsatisfactory in the "tail" region, so that unreliable inter-molecular effects are usually to be expected at intermediate and large R values.

VI.6. Dependence of the First-Order Energy on the Wavefunction Quality.

The remark at the end of the last section suggests that we should try to better appreciate the influence of the quality of the "monomer" unperturbed wavefunctions on the evaluated value of the 1-st order interaction energy. Since the most frequently employed SCF wavefunctions approach only in an imperfect way the H.F. limit (for a good survey of so-called rigorous results, see ref. [106]), it is evident how any calculated 1-st order interaction energy in general reflects inadequacies of the charge density distribution related to the lack of electron correlation and the imperfectly reached H.F. level as well.

We start by considering the effect of progressively augmenting the extension of the basis set, a procedure by which one usually improves the quality of SCF wavefunctions toward the H.F. limit. In Table VI.4 the 1-st order energy for the interaction between two He atoms is examined at two typical distances: R = 5.5 a.u., in the zone of the Van der Waals minimum and R = 7 a.u., where the overlap between the wave-

Table VI.4. 1-st order energy for the interaction He...He as evaluated by progressively better SCF wavefunctions.

n	1	2	4	5
R = 5.5 a.u.				
$-E_{01}$ $(\times 10^5)$ a.u.	0.123	0.603	0.635	0.639
$E_{01,ex}$ $(\times 10^5)$ a.u.	0.757	4.141	4.705	4.566
Δ $(\times 10^5)$ a.u.	1.361	0.332	-0.101	0.003
Total $(\times 10^5)$ a.u.	1.995	3.870	3.969	3.930
R = 7.0 a.u.				
$-E_{01}$ $(\times 10^6)$ a.u.	-0.01	0.07	0.11	0.11
$E_{01,ex}$ $(\times 10^6)$ a.u.	0.09	0.88	1.41	1.12
Δ $(\times 10^6)$ a.u.	0.23	0.16	-0.28	-0.03
Total $(\times 10^6)$ a.u.	0.33	0.97	1.02	0.98

functions of the interacting partners is very small. In the Table, n stands for the number of (Slater-type) orbitals employed for representing the SCF wavefunction of each He atom [157]. E_{01} and $E_{01,ex}$ are the usual Coulomb and exchange contributions to the interaction energy, while the quantity Δ, which we have not met with previously, arises from a definition of the interaction energy different from that we have used, founded on the <u>approximate</u> value of the separate partner energy $\bar{E}_0^A + \bar{E}_0^B = \langle\phi_0^A|\hat{H}_0^A|\phi_0^A\rangle + \langle\phi_0^B|\hat{H}_0^B|\phi_0^B\rangle$ (an upper bound to the true energy). Δ, which has been named "complementary exchange" term [6], has the following expression

$$\Delta = \frac{\langle\phi_0^A\phi_0^B|(\hat{H}_0^A+\hat{H}_0^B-\bar{E}_0^A-\bar{E}_0^B)\hat{\mathcal{P}}^{AB}|\phi_0^A\phi_0^B\rangle}{1 + \langle\phi_0^A\phi_0^B|\hat{\mathcal{P}}^{AB}|\phi_0^A\phi_0^B\rangle}$$

as one easily finds by resorting to a variational computation of the interaction energy in terms of the Heitler-London wavefunction $|\tilde{\Psi}\rangle = \hat{A}_{int}|\phi_0^A\phi_0^B\rangle$, with $\hat{A}_{int}$ given by eq. (V.3.2').

The Table inspection reveals how a SCF wavefunction built up from a minimal basis set is very inadequate for representing both E_{01} and $E_{01,ex}$ while a double-ζ basis representation appears endowed with very gratifying characteristics. It is not unexpected on an intuitive basis that the "complementary exchange" term becomes smaller and smaller as n increases: actually it can be shown that $\Delta=0$ for $|\phi_0^A\rangle$ and $|\phi_0^B\rangle$ the true H.F. eigenstates and $\hat{\mathcal{P}}^{AB} = \hat{P}_{(1)}^{AB}$ (i.e., in the approximation of neglecting multiple exchange effects), as a consequence of the Brillouin's theorem [6]. These results are confirmed by parallel calculations of Snook and Spurling [41] (see next Table).

Some feeling about the importance of including correlation effects in the unperturbed wavefunctions can be gained from the results collected in Table VI.5, which refer again to the interaction He..He [41]. E_{01}^Q represents the global first-order interaction energy evaluated by a 3-term SCF wavefunction for the He atom, while $\delta E_{corr}^{(b)}$ and $\delta E_{corr}^{(c)}$ are changes to such energy produced by differently correlated wavefunctions. $\delta E_{corr}^{(b)}$ corresponds to a wavefunction which included 90% of the radial correlation energy, while $\delta E_{corr}^{(c)}$ arises from a wavefunction

Table VI.5. Correlation correction to H.F. results for the interaction He...He.

R(a.u.)	E_{01}^{Q} $(\times 10^5)$ (a.u.) (a)	δE_{corr} $(\times 10^5)$ (a.u.) (b)	δE_{corr} $(\times 10^5)$ (a.u.) (c)
4.0	147.01	-3.750	-0.416
5.0	13.358	-0.339	-0.024
5.5	3.952	-0.098	-0.005
5.7	2.241	-0.059	-0.003
6.0	1.159	-0.028	-0.001
7.0	0.097	-0.002	-0.000

(a) From a three-term SCF wavefunction.
(b) Radially correlated wavefunction.
(c) Angularly correlated wavefunction.

which included 85% of the angular correlation energy. The rather small effects on the 1-st order interaction energy from intra-correlation corrections should be noted. Even though we have not several well documented examples for deciding that this occurrence is to be expected also for larger systems, we are reasonably confident that SCF wavefunctions of good quality are adequate for calculating E_{01}^{Q}. As a further comment, it appears that the Coulomb energy E_{01} should be less sensitive to correlation corrections than $E_{01,ex}$. In fact, the charge density $\langle\hat{\rho}(\underline{r})\rangle_0$ at any point $\underline{r}$ can be expressed in the form $\langle\hat{\rho}(\underline{r})\rangle_{HF}$ + $\langle\delta\hat{\rho}(\underline{r})\rangle$, where $\langle\delta\hat{\rho}(\underline{r})\rangle$ is the change with respect to the H.F. value caused by the correlation, which is correct through the 1-st order in the intra-correlation effects as a consequence of the Brillouin's theorem. Since $E_{01} = \int d\underline{r}\, d\underline{r}' \langle\hat{\rho}_A(\underline{r})\rangle_0 \langle\hat{\rho}_B(\underline{r}')\rangle_0 / (|\underline{r}'-\underline{r}|)$ (see eq. (VI.1.5)), it follows that $E_{01} = O(\hat{U}^2)$, $\hat{U}$ being the intramonomer perturbation operator (eq. (V.3.8)). An anologous argument does not hold good for $E_{01,ex}$.

VII. THE SECOND-ORDER CONTRIBUTION TO THE INTERACTION ENERGY.

VII.1. Second-order Polarization Energy in the Approximation of Neglecting Charge Overlap Effects.

After having learnt in the last chapter how to evaluate the first-order HL energy, our attention will be turned to the next order contribution to the interaction energy. The importance of making allowance for the 2-nd order contribution to the interaction energy should have clearly been recognized on the basis of our analysis of the system $H...H^+$ in Chap. V. The inspection of Table V.1, in fact, shows that the first-order HL energy $\varepsilon(1)$ for the state $1s\sigma_g$ of H_2^+ exhibits, as a function of the internuclear separation R, a minimum at the approximately correct distance, but deviates considerably from the exact interaction energy value over the whole R range, notably at rather large distances. In the case of the state $2p\sigma_u$, whose behaviour is of particular relevance for us because it mimics the expected behaviour of interacting many-electron partners along the lowest potential energy surface, $\varepsilon(1)$ fails to reproduce the presence of the (shallow) minimum at $R \simeq 12.5$ a.u. and the (weak) attractive region from such R value to larger distances, both of which characteristics require the 2-nd order polarization contribution to be taken into account.

The evaluation of 2-nd order terms in perturbation theory has stirred up a lot of interest in the past years: for this reason, we must be prepared to get involved in a rather long chapter, where we shall meet with different computational approaches, either of a semiempirical or ab-initio character.

From the point of view of our presentation, due to the fact that the 2-nd order contribution to the interaction energy has frequently been identified with the polarization term E_{02} in the approximation of neglecting charge overlap effects, we shall begin our review by reserving the priority tu such subject.

The starting point for the evaluation of the 2-nd order polarization energy E_{02}, which we recall to consist of the sum of the induction and dispersion energy, involves eq. (V.3.15) in conjunction with the use of the multipole expansion for the interaction potential energy operator. The expression for the 2-nd order interaction energy of the molecule A, eq. (V.3.17), can be obtained in a convenient way in two steps. Firstly we observe that

$$\hat{V}^A_{en} + \langle\varphi^B_0|\hat{V}^{AB}_{ee}|\varphi^B_0\rangle = \iint d\underline{r}\, d\underline{r}' \frac{\hat{\rho}^{el}_A(\underline{r})}{|\underline{r}'-\underline{r}|}[\hat{\rho}^{nuc}_B(\underline{r}')+\langle\hat{\rho}^{el}_B(\underline{r}')\rangle_0] \qquad \text{(VII.1.1)}$$

as easily recognized if use is made of the expressions for the nuclear and electronic charge density operator introduced in eq. (VI.1.1); in eq. (VII.1.1) $\langle\hat{\rho}^{el}_B(\underline{r}')\rangle_0 \equiv \langle\varphi^B_0|\hat{\rho}^{el}_B(\underline{r}')|\varphi^B_0\rangle$, the charge density value at the point $\underline{r}'$ associated with the electronic distribution of the molecule B. We then carry out the multipole expansion of $|\underline{r}-\underline{r}'|^{-1}$ according to eq. (VI.3.1); if the space-fixed (parallel) coordinate frames placed in the two molecules are chosen with their z-axes oriented along the intermolecular vector $\underline{R}$ (see Fig. VI.4), the appropriate expression for the $V_{L_aL_b}$ expansion coefficients is that in eq. (VI.3.5), so that

$$\hat{V}_{en} + \langle\varphi^B_0|\hat{V}^{AB}_{ee}|\varphi^B_0\rangle = 4\pi \sum_{L_aL_b} \sum_M \frac{(-1)^{L_b}(L_a+L_b)!}{[(2L_a+1)(2L_b+1)]^{1/2}}$$

$$\times \frac{R^{-(L_a+L_b+1)}}{[(L_a-M)!(L_a+M)!(L_b-M)!(L_b+M)!]^{1/2}}\int d\underline{r}_a \mathcal{Y}^M_{L_a}(\underline{r}_a)\hat{\rho}^{el}_A(\underline{r}_a)$$

$$\times \int d\underline{r}_b \mathcal{Y}^{-M}_{L_b}(\underline{r}_b)\langle\hat{\rho}_B(\underline{r}_b)\rangle_0 \qquad \text{(VII.1.2)}$$

where

$$\langle\hat{\rho}_B(\underline{r}_b)\rangle_0 = \hat{\rho}^{nuc}_B(\underline{r}_b) + \langle\hat{\rho}^{el}_B(\underline{r}_b)\rangle_0 \qquad \text{(VII.1.3)}$$

the total charge density value at $\underline{r}_b$. The quantum numbers that label the sums involved in eq. (VII.1.2) run over the same range as in eq. (VI.2.7).

For the next manipulations, it may be convenient to rewrite eq. (VII.1.2) in the form

$$\hat{V}^A_{en} + \langle \varphi^B_0 | \hat{V}^{AB}_{ee} | \varphi^B_0 \rangle = 4\pi \sum_{L_a L_b} \sum_M (-1)^{L_a} R^{-(L_a+L_b+1)} \left[\frac{(2L_a+2L_b+1)!}{(2L_a+1)!(2L_b+1)!}\right]^{1/2}$$

$$\cdot \begin{pmatrix} L_a & L_b & L_a+L_b \\ M & -M & 0 \end{pmatrix} \int d\underline{r}_a \mathcal{Y}^M_{L_a}(\underline{r}_a) \hat{\rho}^{el}_A(\underline{r}_a) \int d\underline{r}_b \mathcal{Y}^{-M}_{L_b}(\underline{r}_b) \langle \hat{\rho}_B(\underline{r}_b) \rangle_0 \qquad \text{(VII.1.4)}$$

in terms of appropriate Wigner 3-j coefficients $\begin{pmatrix} L_a & L_b & L_a+L_b \\ M & -M & 0 \end{pmatrix}$ [143].

In order to take advantage of the molecular symmetry and to put in explicit evidence the dependence on the orientation of the interacting "monomers", we introduce body-fixed coordinate axes (see Fig. VI.4) and make use of the relationship between space- and body-fixed spherical harmonics, eq. (VI.2.8); it is straightforward to verify that eq. (VII.1.4) becomes

$$\hat{V}^A_{en} + \langle \varphi^A_0 | \hat{V}^{AB}_{ee} | \varphi^A_0 \rangle = \sum_{L_a L_b} \sum_M \sum_{m_a m_b} (-1)^{L_a} R^{-(L_a+L_b+1)} \left[\frac{(2L_a+2L_b+1)!}{(2L_a)!(2L_b)!}\right]^{1/2}$$

$$\times \begin{pmatrix} L_a & L_b & L_a+L_b \\ M & -M & 0 \end{pmatrix} \mathbb{D}^{L_a *}_{M,m_a}(\Omega_a) \mathbb{D}^{L_b *}_{-M,m_b}(\Omega_b) (\hat{Q}^{el}_A)^{m_a}_{L_a} (\hat{Q}_B)^{m_b}_{L_b} \qquad \text{(VII.1.5)}$$

where we have introduced the permanent multipole moment $(Q_B)^{m_b}_{L_b}$ of the "monomer" B,

$$(Q_B)^{m_b}_{L_b} = \left(\frac{4}{2L_b+1}\right)^{1/2} \int d\underline{r}_b \mathcal{Y}^{m_b}_{L_b}(\underline{r}_b) \langle \hat{\rho}_B(\underline{r}_b) \rangle_0 \qquad \text{(VII.1.6)}$$

referred to body-fixed axes in such molecule, and the corresponding operator quantity $(\hat{Q}_A^{el})_{L_a}^{m_a}$ for the electronic component of the charge distribution of the molecule A, an logously referred to body-fixed axes,

$$(\hat{Q}_A^{el})_{L_a}^{m_a} = \left(\frac{4\pi}{2L_a+1}\right)^{1/2} \int d\underline{r}_a \mathcal{Y}_{L_a}^{m_a}(\underline{r}_a)\hat{\rho}_A^{el}(\underline{r}_a) \qquad \text{(VII.1.7)}$$

(Eqs. (VII.1.6) and (VII.1.7) obviously agree with the related eq. (VI. 3.8)).

We are now in a position to express in a convenient way the induction energy $E_{02,ind}^{A}$ of the molecule A in the field of the molecule B; from eqs. (V.3.17) and (VII.1.5), we get the following (complicate) result [158]

$$E_{02,ind}^{A}(R,\underline{\Omega}_a,\underline{\Omega}_b) = \sum_{L_aL_b} \sum_{L_a'L_b'} \sum_{MM'} \sum_{m_am_b} \sum_{m_a'm_b'} (-1)^{L_a+L_a'}$$

$$\times \left[\frac{(2L_a+2L_b+1)!\,(2L_a'+2L_b'+1)!}{(2L_a)!\,(2L_b)!\,(2L_a')!\,(2L_b')!}\right]^{1/2} R^{-(L_a+L_a'+L_b+L_b'+2)} \begin{pmatrix} L_a & L_b & L_a+L_b \\ M & -M & 0 \end{pmatrix}$$

$$\times \begin{pmatrix} L_a' & L_b' & L_a'+L_b' \\ M' & -M' & 0 \end{pmatrix} \mathbb{D}_{M,m_a}^{L_a\,*}(\underline{\Omega}_a)\,\mathbb{D}_{M',m_a'}^{L_a'\,*}(\underline{\Omega}_a)\,\mathbb{D}_{-M,m_b}^{L_b\,*}(\underline{\Omega}_b)\,\mathbb{D}_{-M',m_b'}^{L_b'\,*}(\underline{\Omega}_b)\,(Q_B)_{L_b}^{m_b}$$

$$\times (Q_B)_{L_b'}^{m_b'} \langle \varphi_0^A | (\hat{Q}_A^{el})_{L_a}^{m_a} \hat{R}_0^A (\hat{Q}_A^{el})_{L_a'}^{m_a'} | \varphi_0^A \rangle \qquad \text{(VII.1.8)}$$

for the anisotropic induction energy of the molecule A in the field of the molecule B (the orientation dependence being completely contained in the rotation matrix elements $\mathbb{D}_{M,m}^{L}(\underline{\Omega})$). An entirely analogous result for the induction energy $E_{02,ind}^{B}$ stems from eq. (VII.1.8) by changing A$\rightleftarrows$B, a$\rightleftarrows$b

A more compact and elegant expression for $E_{02,ind}$ (and $E_{02,disp}$) has been derived by Wormer [145] and Wormer *et al.* [159] who have pushed the rotational dependence of E_{02} to its most simplified form;

however we shall content ourselves with the less refined result (VII. 1.8), without becoming involved in irreducible tensor recoupling procedures that are required for achieving such result.

Although recent advances in molecular-beam scattering and dimer spectroscopy make the evaluation of the general anisotropic contributions to the interaction energy quite interesting [160-165], the value of $E_{02,ind}$ averaged over all orientations of the "monomers" appears surely a basic quantity. If we assume that a classical average procedure with equal weights is adequate, from the result [142]

$$\int d\underline{\Omega} \mathbb{D}^{J_1 *}_{\mu_1 m_1}(\underline{\Omega}) \mathbb{D}^{J_2}_{\mu_2 m_2}(\underline{\Omega}) = 8\pi^2 (2J_1+1)^{-1} \delta_{J_1 J_2} \delta_{\mu_1 \mu_2} \delta_{m_1 m_2}$$

and $\mathbb{D}^{J*}_{\mu m}(\underline{\Omega}) = (-1)^{\mu-m} \mathbb{D}^{J}_{-\mu,-m}(\underline{\Omega})$, we find for the rotationally averaged induction energy $(E^A_{02,ind})_{av}$ the expression

$$(E^A_{02,ind})_{av} = \sum_{L_a L_b} \sum_{m_a m_b} (-1)^{m_a+m_b} \frac{(2L_a+2L_b)!}{(2L_a+1)!(2L_b+1)!} R^{-2(L_a+L_b+1)}$$

$$\times (Q_B)^{m_b}_{L_b} (Q_B)^{-m_b}_{L_b} \langle \varphi^A_0 | (\hat{Q}^{el}_A)^{m_a}_{L_a} \hat{R}^A_0 (\hat{Q}^{el}_A)^{-m_a}_{L_a} | \varphi^A_0 \rangle \qquad \text{(VII.1.9)}$$

Eq. (VII.1.9) can be given a much more compact form in terms of the static multipole polarizabilities of the molecule A. For a pair of L, L' values, let us consider the (2L+1)(2L'+1) quantities

$$\alpha^{L',m'}_{L,m} \equiv -2\langle \varphi_0 | (\hat{Q}^{el})^m_L \hat{R}_0 (\hat{Q}^{el})^{m'}_{L'} | \varphi_0 \rangle$$

$$= 2 \sum_{k\neq 0} E^{-1}_{k0} \langle \varphi_0 | (\hat{Q}^{el})^m_L | \varphi_k \rangle \langle \varphi_k | (\hat{Q}^{el})^{m'}_{L'} | \varphi_0 \rangle \qquad \text{(VII.1.10)}$$

where in the last expression, which follows from using the spectral resolution of $\hat{R}_0$, eq. (III.6.6), we have put $E_{k0} = E_k - E_0$, the excitation energy of the system from the unperturbed initial state to the k-th one*. From the tensorial quantities $\alpha_{L,m}^{L',m'}$ we may construct <u>irreducible</u> tensors according to eq. (VI.3.4); in fact, with $\langle Lm;L'm'|JM\rangle$ an appropriate Clebsh-Gordan coefficient,

$$\sum_{mm'} \langle\varphi_0|(\hat{Q}^{el})^m_L|\varphi_k\rangle\langle\varphi_k|(\hat{Q}^{el})^{m'}_{L'}|\varphi_o\rangle\langle Lm;L'm'|JM\rangle$$

$$\equiv [\langle\varphi_o|\hat{Q}^{el}_L|\varphi_k\rangle \otimes \langle\varphi_k|\hat{Q}^{el}_{L'}|\varphi_o\rangle]^J_M$$

represents the coupling of an electric 2^L-pole moment matrix element with an analogous electric $2^{L'}$-pole moment matrix element, to give a 2^J-pole moment. We have therefore [145,165]

$$2\sum_{mm'}\sum_{k\neq 0} E^{-1}_{k0}\langle\varphi_o|(\hat{Q}^{el})^m_L|\varphi_k\rangle\langle\varphi_k|(\hat{Q}^{el})^{m'}_{L'}|\varphi_o\rangle\langle Lm;L'm'|JM\rangle$$

$$= 2\sum_{k\neq 0} E^{-1}_{k0}[\langle\varphi_o|\hat{Q}^{el}_L|\varphi_k\rangle \otimes \langle\varphi_k|\hat{Q}^{el}_{L'}|\varphi_o\rangle]^J_M \equiv \alpha_{(L,L')J,M} \qquad \text{(VII.1.11)}$$

for the J-th rank irreducible polarizability tensor which results from coupling a 2^L-pole with a $2^{L'}$-pole. For the case $J = 0$ we get

$$\alpha_{(L,L')0,0} = (-1)^L(2L+1)^{-\frac{1}{2}}\delta_{LL'}\sum_{m=-L}^{+L}(-1)^m\alpha_{L,m}^{L',-m}$$

$$= -2(-1)^L(2L+1)^{-\frac{1}{2}}\delta_{LL'}\sum_m(-1)^m\langle\varphi_0|(\hat{Q}^{el})^m_L\hat{R}_0(\hat{Q}^{el})^{-m}_L|\varphi_o\rangle \qquad \text{(VII.1.12)}$$

An analogous coupling procedure can be applied to the <u>permanent</u> multipole moments Q_L of the molecule B; from eq. (VI.3.4) we find in a

---

*Our definition of multipole polarizability is essentially the same as that used by Oka in ref. [166], the difference consisting in a mere multiplicative factor (depending on L,L').

straightforward way

$$[Q_L \otimes Q_{L'}]_0^0 = (-1)^L (2L+1)^{-\frac{1}{2}} \delta_{LL'} \sum_{m=-L}^{+L} (-1)^m Q_L^m Q_L^{-m} \qquad \text{(VII.1.13)}$$

By using eqs. (VII.1.12), (VII.1.13), eq. (VII.1.9) can be rewritten in the form

$$(E^A_{02,in})_{av} = -\frac{1}{2} \sum_{L_a L_b} (-1)^{L_a+L_b} \binom{2L_a+2L_b}{2L_a} [(2L_a+1)(2L_b+1)]^{-\frac{1}{2}}$$

$$\times R^{-2(L_a+L_b+1)} \alpha^A_{(L_a,L_a)0,0} [(Q_B)_{L_b} \otimes (Q_B)_{L_b}]_0^0 \qquad \text{(VII.1.14)}$$

The relative simplicity of this expression for the rotationally averaged induction energy should be contrasted with the anisotropic interaction result, eq. (VII.1.8): in that case, in fact, off-diagonal multipole polarizabilities are in general involved, as already remarked [158].

The 2-nd order dispersion energy, eq. (V.3.18), can be cast into a useful form by the same procedure used for $E_{02,ind}$. The starting point is now offered by the following representation of the interelectronic potential energy operator,

$$\hat{V}^{AB}_{ee} = \iint d\underline{r}\, d\underline{r}' \frac{\hat{\rho}^A_{el}(\underline{r}')\hat{\rho}^B_{el}(\underline{r}')}{|\underline{r}-\underline{r}'|}$$

[see eqs. (VI.1.1)-(VI.1.3)], together with a multipole expansion of the Coulomb potential $|\underline{r}-\underline{r}'|^{-1}$. If the involved multipole moments are successively referred to body-fixed axes, one is led to the following expression for the orientation dependent 2-nd order dispersion energy of two arbitrary "monomers" [158]

$$E_{02,disp}(R,\underline{\Omega}_a,\underline{\Omega}_b) = \sum_{L_aL_b} \sum_{L_a'L_b'} \sum_{MM'} \sum_{m_am_b} \sum_{m_a'm_b'} (-1)^{L_a+L_a'}$$

$$\times \left[\frac{(2L_a+2L_b+1)!\,(2L_a'+2L_b'+1)!}{(2L_a)!\,(2L_b)!\,(2L_a')!\,(2L_b')!}\right]^{1/2} R^{-(L_a+L_a'+L_b+L_b'+2)} \begin{pmatrix} L_a & L_b & L_a+L_b \\ M & -M & 0 \end{pmatrix}$$

$$\times \begin{pmatrix} L_a' & L_b' & L_a'+L_b' \\ M' & -M' & 0 \end{pmatrix} \mathbb{D}^{L_a*}_{M,m_a}(\underline{\Omega}_a)\, \mathbb{D}^{L_a'*}_{M',m_a'}(\underline{\Omega}_a)\, \mathbb{D}^{L_b*}_{-M,m_b}(\underline{\Omega}_b)\, \mathbb{D}^{L_b'*}_{-M',m_b'}(\underline{\Omega}_b)$$

$$\times \langle\varphi_o^A\varphi_0^B|\,(\hat{Q}_A^{el})^{m_a}_{L_a}(\hat{Q}_B^{el})^{m_b}_{L_b}\,\hat{R}_0^{AB}\,(\hat{Q}_A^{el})^{m_a'}_{L_a'}(\hat{Q}_B^{el})^{m_b'}_{L_b'}|\varphi_o^A\varphi_0^B\rangle \qquad \text{(VII.1.15)}$$

$\hat{R}_0^{AB}$ being the resolvent defined by eq. (V.3.11). Some formal similarity between eq. (VII.1.15) and eq. (VII.1.8) for the induction energy is apparent; in particular, the orientation dependence is also in the present case contained in the elements of the irreducible representation matrices $\mathbb{D}^{L}_{M,m}$ of the rotation group. If one is satisfied with the value of the dispersion energy averaged with respect to all orientations of the partners, resorting to the same manipulations used for the preceding treatment of the induction energy leads in a rather straightforward way to

$$(E_{02,disp})_{av} = \sum_{L_aL_b} \sum_{m_am_b} (-1)^{m_a+m_b} \frac{(2L_a+2L_b)!}{(2L_a+1)!\,(2L_b+1)!} R^{-2(L_a+L_b+1)}$$

$$\times \langle\varphi_0^A\varphi_0^A|\,(\hat{Q}_A^{el})^{m_a}_{L_a}\,(\hat{Q}_B^{el})^{m_b}_{L_b}\,\hat{R}_0^{AB}\,(\hat{Q}_A^{el})^{-m_a}_{L_a}(\hat{Q}_B^{el})^{-m_b}_{L_b}|\varphi_o^A\varphi_0^B\rangle \qquad \text{(VII.1.16)}$$

a function of the intermolecular distance R only.

Contrary to the case of the induction energy (and in accordance with its interpretation put forward in sec. V.3), the dispersion energy in the multipole approximation does not involve permanent multipole moments, the entire effect being contained in the matrix element $\langle\varphi_0^A\varphi_0^B|(\hat{Q}_A^{el})_{L_a}^{m_a}(\hat{Q}_B^{el})_{L_b}^{m_b}\hat{R}_0^{AB}(\hat{Q}_A^{el})_{L_a}^{-m_a}(\hat{Q}_B^{el})_{L_b}^{-m_b}|\varphi_0^A\varphi_0^B\rangle$. We see therefore that in the approximation of neglecting charge overlap effects the induction energy contribution to the interaction energy can be missing (when the interacting partners do not possess permanent multipole moments), while the dispersion energy is always a not vanishing quantity.

The matrix element which determines the dispersion energy can be expressed in terms of the eigenstates to the unperturbed Hamiltonian operator $\hat{H}_0 = \hat{H}_0^A+\hat{H}_0^B$; from the expression for the resolvent $\hat{R}_0^{AB}$, eq. (V.3.11), we obtain

$$\langle\varphi_0^A\varphi_0^B|(\hat{Q}_A^{el})_{L_a}^{m_a}(\hat{Q}_B^{el})_{L_b}^{m_b}\hat{R}_0^{AB}(\hat{Q}_A^{el})_{L_a'}^{m_a'}(\hat{Q}_B^{el})_{L_b'}^{m_b'}|\varphi_0^A\varphi_0^B\rangle$$

$$=-\sum_{j\neq 0}\sum_{k\neq 0}\frac{\langle\varphi_0^A|(\hat{Q}_A^{el})_{L_a}^{m_a}|\varphi_j^A\rangle\langle\varphi_j^A|(\hat{Q}_A^{el})_{L_a'}^{m_a'}|\varphi_0^A\rangle\langle\varphi_0^B|(\hat{Q}_B^{el})_{L_b}^{m_b}|\varphi_k^B\rangle\langle\varphi_k^B|(\hat{Q}_B^{el})_{L_b'}^{m_b'}|\varphi_0^B\rangle}{E_{j0}^A+E_{k0}^B}$$

(VII.1.17)

where we have put $E_{j0} = E_j-E_0$, the excitation energy from the state $|\varphi_0\rangle$ to the state $|\varphi_j\rangle$ and the sums run over the complete sets of eigenstates to $\hat{H}_0^A$ and $\hat{H}_0^B$. Apart from the problems to be encountered in the actual evaluation of eq. (VII.1.17), we should note its non-separable "two-center" form, caused by the presence of the denominator $E_{j0}^A+E_{k0}^B$. This is at variance with the case of the induction energy, where a natural factorization of the interaction matrix element occurred.

A very interesting transformation that leads to a factorized form of the preceding result, eq. (VII.1.17), is however possible, even though at the cost of different complications. From the following integral identity (easily verified, for example, by contour integration)

$$(a+b)^{-1} = (2/\pi)\int_0^\infty dx\,\frac{ab}{(a^2+x^2)(b^2+x^2)} \qquad (a>0,\ b>0) \qquad \text{(VII.1.18)}$$

for molecules both in their ground state eq. (VII.1.17) can be rewritten as

$$\langle \varphi_0^A \varphi_0^B | (\hat{Q}_A^{el})_{L_a}^{m_a} (\hat{Q}_B^{el})_{L_b}^{m_b} \hat{R}_0^{AB} (\hat{Q}_A^{el})_{L'_a}^{m'_a} (\hat{Q}_B^{el})_{L'_b}^{m'_b} | \varphi_0^A \varphi_0^B \rangle$$

$$= -\frac{2}{\pi} \int_0^\infty d\omega \left[\sum_{j\neq 0} \frac{E_{j0}^A \langle \varphi_0^A | (\hat{Q}_A^{el})_{L_a}^{m_a} | \varphi_j^A \rangle \langle \varphi_j^A | (\hat{Q}_A^{el})_{L'_a}^{m'_a} | \varphi_0^A \rangle}{(E_{j0}^A)^2 + \omega^2} \right]$$

$$\times \left[\sum_{k\neq 0} \frac{E_{k0}^B \langle \varphi_0^B | (\hat{Q}_B^{el})_{L_b}^{m_b} | \varphi_k^B \rangle \langle \varphi_k^B | (\hat{Q}_B^{el})_{L'_b}^{m'_b} | \varphi_0^B \rangle}{(E_{k0}^B)^2 + \omega^2} \right]$$

$$\equiv -\frac{1}{2\pi} \int_0^\infty d\omega \; \alpha_{L_a,m_a}^{L'_a,m'_a}(i\omega) \; \alpha_{L_b,m_b}^{L'_b,m'_b}(i\omega) \qquad \text{(VII.1.19)}$$

where $\alpha(i\omega)$ is the value of the dynamic electric polarizability $\alpha(\omega)$ prolonged to points of the imaginary axis [167]

$$\alpha_{L,m}^{L',m'}(\omega) = 2 \sum_{k\neq 0} \frac{E_{k0} \langle \varphi_0 | (\hat{Q}^{el})_L^m | \varphi_k \rangle \langle \varphi_k | (\hat{Q}^{el})_{L'}^{m'} | \varphi_0 \rangle}{E_{k0}^2 - \omega^2} \qquad \text{(VII.1.20)}$$

If methods for the evaluation of dynamic polarizabilities and their extension in the complex plane to the imaginary axis can be devised, we have at our disposal an elegant and powerful way for obtaining dispersion energies.

If we make use of eq. (VII.1.19) and extend to $\alpha_{L,m}^{L',-m}(i\omega)$ the coupling procedure expressed by eq. (VII.1.12), so as to generate the scalar $\alpha^0_{(L,L')J=0}(i\omega)$, we may rewrite eq. (VII.1.16) for the orientationally averaged dispersion energy between two arbitrary partners in the form

$$(E_{02,\text{disp}})\text{av} = -\frac{1}{2\pi} \sum_{L_a L_b} (-1)^{L_a+L_b} \binom{2L_a+2L_b}{2L_a} \left[(2L_a+1)(2L_b+1)\right]^{-\frac{1}{2}}$$

$$\times R^{-2(L_a+L_b+1)} \int_0^{\infty} d\omega\, \alpha^{A}_{(L_a,L_a)0,0}(i\omega)\, \alpha^{B}_{(L_b,L_b)0,0}(i\omega) \qquad \text{(VII.1.21)}$$

which should be contrasted with eq. (VII.1.14) for the averaged induction energy. This expression for the dispersion energy is frequently referred to as Casimir-Polder (integral) formula*; the appellation "one-center method" is also employed to characterize the computational approach along these lines [2,144], just to emphasize the reduction of the original two-molecule problem into two single-molecule problems.

The expression for the induction and dispersion energy previously obtained are most frequently put in the following form

$$E_{02,ind} = - \sum_{m=6}^{\infty} R^{-m} C_m^{ind}$$

$$\text{(VII.1.22)}$$

$$E_{02,disp} = - \sum_{m=6}^{\infty} R^{-m} C_m^{disp}$$

---

*Eqs. (VII.1.15) and (VII.1.21) predict for the dispersion energy a dependence on the intermolecular distance R which at the lowest order varies as R^{-6}. At very large distances, however, deviations from such behaviour become manifest, which in conclusion can be attributed to the finite time required for an electromagnetic perturbation to be transmitted from one molecule to the other (because of the finite velocity of light). The result is that at such separations the dispersion energy at the lowest order varies as R^{-7}, as first found by Casimir and Polder [168]. By using a relativistic treatment, these authors were led in a natural way to an expression for the dispersion energy involving an integration similar to that occurring in eq. (VII.1.21) (which, in our case, is admittedly the result of an ingenious trick). Several authors in various papers and reviews have subsequently re-examined the Casimir and Polder's conclusions [169-176] (in some papers [177-180] there are also suggested suitable techniques for the computation of retarded Van der Waals coefficients). As far as we are concerned, although the subject is quite interesting, we have decided to leave it out from the present notes.

which define the induction and dispersion Van der Waals interaction coefficients C_m^{ind} and C_m^{disp}, respectively. Eqs. (VII.1.22) correspond clearly to another way of regrouping the terms of the series for the induction and dispersion energies, so as to put in evidence only their structure of series in powers of R^{-1}. These interaction coefficients are independent of the intermolecular distance R, but in the anisotropic case are obviously functions of the relative orientation of the two interacting partners.

Before ending this section, we think it convenient to emphasize that the series (VII.1.22) are very likely asymptotic in nature [58, 97], as for the interaction $H...H^+$ between an hydrogen atom and a proton [92] (see sec. V.2). We are not aware that a general proof of the asymptotic behaviour of the series (VII.1.22) has been put forward, contrary to the case of the over-all interaction energy, for which an elegant demonstration is available [72]: our slightly cautious statement about the character of the series for both $E_{02,ind}$ and $E_{02,disp}$ reflects only this lack of a general rigorous justification*.

VII.2. Induction and Dispersion Coefficients for Two Interacting Linear Molecules.

Contrary to the orientationally averaged induction and dispersion energy expressions, which are fairly simple [see eqs. (VII.1.14), (VII.1.16), (VII.1.21)], the general expressions for the corresponding anisotropic energies, eqs. (VII.1.8) and (VII.1.15), appear really formidable. Noticeable troubles of formal character are to be expected when the series is pushed beyond the first term, so much so that resorting to a computer program is likely the most profitable way for generating correct formulae for the various contributions.

Just to give an example of application of the expressions for $E_{02,ind}$ and $E_{02,disp}$ deduced in the preceding section, we shall con-

---

*In the past few years the R^{-1} expansion for the 2-nd order interaction between two ground state hydrogen atoms has been shown to diverge for all R values; however its asymptotic character, even though plausible, has not been explicitly demonstrated [181].

sider the case of two linear molecules; after specializing to this situation the general results, we shall confine for simplicity our attention to the lowest contribution to both $E_{02,ind}$ and $E_{02,disp}$.

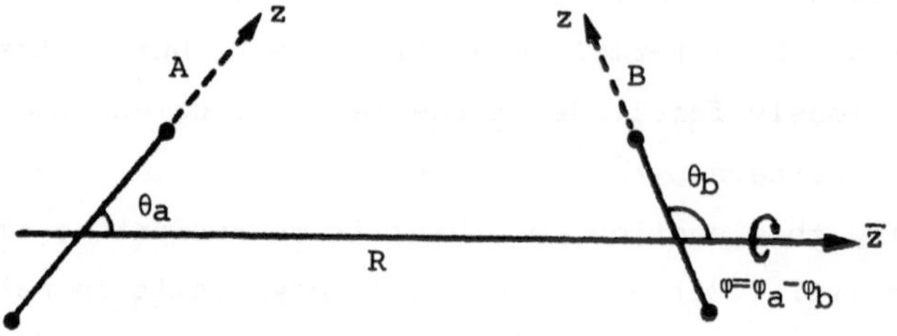

Fig. VII.1. Configuration of two interacting linear molecules A, B, along with the coordinates $R,\theta_a,\theta_b,\varphi$ used for its characterization.

We shall start with the induction energy. For linear molecules, if we choose the body-fixed coordinate system with the z-axis oriented along the internuclear vector (see Fig. VII.1), we have

$$Q_L^m = Q_L\,\delta_{m0}$$

$$\alpha_{L,m}^{L',m'} = \alpha_{L,m}^{L',-m}\,\delta_{m,-m'} \qquad \text{(VII.2.1)}$$

(see also sects. VI.2 and VI.4), so that eq. (VII.1.8) can be written as

$$E_{02,ind}^{A}(R,\underline{\Omega}_a,\underline{\Omega}_b) = -\frac{1}{2}\sum_{L_aL_b}\sum_{L_a'L_b'}\sum_{MM'}\sum_{m_a}(-1)^{L_a+L_b'}\left[\frac{(2L_a+2L_b+1)!}{(2L_a)!(2L_b)!}\right]^{1/2}$$

$$\times\left[\frac{(2L_a'+2L_b'+1)!}{(2L_a')!(2L_b')!}\right]^{1/2} R^{-(L_a+L_b+L_a'+L_b'+2)}\begin{pmatrix}L_a & L_b & L_a+L_b\\ M & -M & 0\end{pmatrix}\begin{pmatrix}L_a' & L_b' & L_a'+L_b'\\ M' & -M' & 0\end{pmatrix}$$

$$\times\, Q_{L_b} Q_{L_b'}\, \alpha^{L_a',-m_a}_{L_a,m_a}\, \mathbb{D}^{L_a *}_{M,m_a}(\underline{\Omega}_a)\, \mathbb{D}^{L_a' *}_{M',-m_a}(\underline{\Omega}_a)\, \mathbb{D}^{L_b *}_{-M,0}(\underline{\Omega}_b)\, \mathbb{D}^{L_b' *}_{-M',0}(\underline{\Omega}_b) \qquad \text{(VII.2.2)}$$

Making use of the Clebsch-Gordan reduction formula for the direct product $\mathbb{D}^{L} \otimes \mathbb{D}^{L'}$ [143,165]

$$\mathbb{D}^{L}_{M,m}(\underline{\Omega})\,\mathbb{D}^{L'}_{M',-m}(\underline{\Omega}) =$$

$$= \sum_{J=|L-L'|}^{J=L+L'} (-1)^{M+M'} (2J+1) \begin{pmatrix} L & L' & J \\ M & M' & -(M+M') \end{pmatrix} \begin{pmatrix} L & L' & J \\ m & -m & 0 \end{pmatrix} \mathbb{D}^{J}_{M+M',0}(\underline{\Omega}) \qquad \text{(VII.2.3)}$$

eq. (VII.2.2) becomes

$$E^{A}_{02,\mathrm{ind}}(R,\underline{\Omega}_a,\underline{\Omega}_b) = -\frac{1}{2} \sum_{L_a L_b} \sum_{L_a' L_b'} \sum_{MM'} \sum_{m_a} \sum_{J_a J_b} (-1)^{L_a+L_b}$$

$$\times \left[\frac{(2L_a+2L_b+1)!\,(2L_a'+2L_b'+1)!}{(2L_a)!\,(2L_b)!\,(2L_a')!\,(2L_b')!}\right]^{1/2} R^{-(L_a+L_a'+L_b+L_b'+2)} (2J_a+1)(2J_b+1)$$

$$\times\, Q_{L_b} Q_{L_b'}\, \alpha^{L_a',-m_a}_{L_a,m_a} \begin{pmatrix} L_a & L_b & L_a+L_b \\ M & -M & 0 \end{pmatrix} \begin{pmatrix} L_a' & L_b' & L_a'+L_b' \\ M' & -M' & 0 \end{pmatrix} \begin{pmatrix} L_a & L_a' & J_a \\ M & M' & -(M+M') \end{pmatrix} \begin{pmatrix} L_a & L_a' & J_a \\ m_a & -m_a & 0 \end{pmatrix}$$

$$\times \begin{pmatrix} L_b & L_b' & J_b \\ -M & -M' & M+M' \end{pmatrix} \begin{pmatrix} L_b & L_b' & J_b \\ 0 & 0 & 0 \end{pmatrix} \mathbb{D}^{J_a *}_{M+M',0}(\underline{\Omega}_a)\, \mathbb{D}^{J_b *}_{-(M+M'),0}(\underline{\Omega}_b) \qquad \text{(VII.2.4)}$$

It is now convenient to introduce the zero-th component of the J-th rank irreducible polarizability tensor, eq. (VII.1.11), which in terms of Wigner 3-J coefficients is expressed as

$$\alpha_{(L,L')J,0} = (-1)^{L-L'} (2J+1)^{1/2} \sum_{m} \begin{pmatrix} L & L' & J \\ m & -m & 0 \end{pmatrix} \alpha^{L',-m}_{L,m} \qquad \text{(VII.2.5)}$$

Eq. (VII.2.4) then assumes the following final form

$$E^{A}_{02,\mathrm{ind}}(R,\underline{\Omega}_a,\underline{\Omega}_b) = -2\pi \sum_{L_aL_b} \sum_{M} \sum_{L'_aL'_b} \sum_{M'} \left[\frac{(2L_a+2L_b+1)!\,(2L'_a+2L'_b+1)!}{(2L_a)!\,(2L_b)!\,(2L'_a)!\,(2L'_b)!}\right]^{1/2}$$

$$\times\, R^{-(L_a+L_b+L'_a+L'_b+2)} Q_{L_b} Q_{L'_b} \begin{pmatrix} L_a & L_b & L_a+L_b \\ M & -M & 0 \end{pmatrix} \begin{pmatrix} L'_a & L'_b & L'_a+L'_b \\ M' & -M' & 0 \end{pmatrix}$$

$$\times \sum_{J_a} \alpha_{(L_a,L'_a)J_a,0} \begin{pmatrix} L_a & L'_a & J_a \\ M & M' & -(M+M') \end{pmatrix} Y^{M+M'}_{J_a}(\theta_a,\varphi_a) \sum_{J_b} (2J_b+1)^{1/2}$$

$$\times \begin{pmatrix} L_b & L'_b & J_b \\ -M & -M' & M+M' \end{pmatrix} \begin{pmatrix} L_b & L'_b & J_b \\ 0 & 0 & 0 \end{pmatrix} Y^{-(M+M')}_{J_b}(\theta_b,\varphi_b) \qquad \text{(VII.2.6)}$$

where we have used $\mathbf{D}^{J}_{m,0}(\underline{\Omega}) \equiv \mathbf{D}^{J}_{m,0}(\varphi,\theta,0) = (4\pi/2J+1)^{\frac{1}{2}} Y^{M*}_{J}(\theta,\varphi)$ [142,143]. It should be clear that the permanent multipole moments Q_{L_b} refer to the molecule B, while $\alpha_{(L_a,L'_a)J_a,0}$ denote polarizabilities of the molecule A.

The lowest order contribution to $E_{02,\mathrm{ind}}$ stems from $L_a=L_b=L'_a=L'_b=1$ and varies with the internuclear distance as R^{-6}; the corresponding interaction coeffcient $C^{\mathrm{ind}(A)}_{6}$ [see eq. (VII.1.22)] is given by

$$C^{\mathrm{ind}(A)}_{6}(\underline{\Omega}_a,\underline{\Omega}_b) = \frac{2\pi\cdot 5!}{4} [Q_{1B}]^2 \sum_{M=-1}^{+1} \sum_{M'=-1}^{+1} \begin{pmatrix} 1 & 1 & 2 \\ M & -M & 0 \end{pmatrix} \begin{pmatrix} 1 & 1 & 2 \\ M' & -M' & 0 \end{pmatrix}$$

$$\times \sum_{J_a} \begin{pmatrix} 1 & 1 & J_a \\ M & M' & -(M+M') \end{pmatrix} \alpha_{(11)J_a,0} Y^{M+M'}_{J_a}(\theta_a,\varphi_a)$$

$$\times \sum_{J_b} (2J_b+1)^{\frac{1}{2}} \begin{pmatrix} 1 & 1 & J_b \\ -M & -M' & M+M' \end{pmatrix} \begin{pmatrix} 1 & 1 & J_b \\ 0 & 0 & 0 \end{pmatrix} Y^{-(M+M')}_{J_b}(\theta_b,\varphi_b) \qquad \text{(VII.2.7)}$$

The analogous coefficient $C^{\mathrm{ind}(B)}_{6}$, which represents the effect on the molecule B induced by the permanent dipole moment of the molecule A, is simply obtained from eq. (VII.2.7) by exchanging $A \leftrightarrows B$, $a \leftrightarrows b$.

The next manipulations of eq. (VII.2.7) require some (annoying) algebra based on the properties of the 3-J coefficients. The irreducible polarizabilities involved, $\alpha_{(1,1)2,0}$ and $\alpha_{(1,1)2,0}$, follow from eq. (VII.2.5): $\alpha_{(1,1)2,0} = (6)^{-\frac{1}{2}}[\alpha^{1,1}_{1,-1} + \alpha^{1,-1}_{1,1} + 2\alpha^{1,0}_{1,0}]$, $\alpha_{(1,1)0,0} = -(3)^{-\frac{1}{2}} \cdot [\alpha^{1,0}_{1,0} - \alpha^{1,-1}_{1,1} - \alpha^{1,1}_{1,-1}]$. In order to gain a better appreciation of these quantities, however, it may be interesting to express them in terms of <u>Cartesian tensor polarizabilities</u> [2,6,182]; it is rather straightforward to verify that

$$\alpha_{(1,1)2,0} = \begin{cases} (6)^{-\frac{1}{2}}[2\alpha_{zz} - \alpha_{xx} - \alpha_{yy}] & \text{(any molecule)} \\ 2(6)^{-\frac{1}{2}}[\alpha^{\parallel} - \alpha^{\perp}] & \text{(axial molecule)} \end{cases} \tag{VII.2.8}$$

$$\alpha_{(1,1)0,0} = \begin{cases} -(3)^{-\frac{1}{2}}[\alpha_{xx} + \alpha_{yy} + \alpha_{zz}] & \text{(any molecule)} \\ -(3)^{-\frac{1}{2}}[2\alpha^{\perp} + \alpha^{\parallel}] & \text{(axial molecule)} \end{cases}$$

where, for instance, the xx-component of the dipole polarizability tensor is defined as

$$\alpha_{xx} = -2\langle\varphi_0|\hat{\mu}_x \hat{R}_0 \hat{\mu}_x|\varphi_0\rangle \tag{VII.2.9}$$

$\hat{\mu}_x = \int d\underline{r}\; x\; \hat{\rho}^{el}(\underline{r})$ being the x-component of the dipole moment operator $\hat{\mu}$, (see sec. VI.3). For an axial molecule, with axes so chosen that z lies along the internuclear axis, $\alpha_{xx} = \alpha_{yy} \equiv \alpha^{\perp}$ and $\alpha_{zz} \equiv \alpha^{\parallel}$. We see therefore that $\alpha_{(1,1)0,0}$ is proportional to the so-called <u>isotropic part</u> $\bar{\alpha}$ of the tensor α_{ij},

$$\bar{\alpha} = (3)^{-1}[\alpha_{xx} + \alpha_{yy} + \alpha_{zz}] \tag{VII.2.10}$$

while $\alpha_{(1,1)2,0}$ is proportional to its <u>anisotropy</u> γ,

$$\gamma = \alpha^{\parallel} - \alpha^{\perp} \qquad \text{(VII.2.11)}$$

If we put $\mu_B \equiv Q_{1B}$, the permanent dipole moment of the molecule B, we find after some algebra

$$C_6^{ind(A)}(\theta_a,\theta_b,\varphi) = \mu_B^2\left[\frac{1}{2}\,\bar{\alpha}_A(3\cos^2\theta_b+1)+\frac{1}{2}(12\cos^2\theta_a\cos^2\theta_b+3\mathrm{sen}^2\theta_a\cdot\right.$$

$$\left.\cdot\mathrm{sen}^2\theta_b\cos^2\varphi-3\cos^2\theta_b-1-12\,\mathrm{sen}\theta_a\cos\theta_a\mathrm{sen}\theta_b\cos\theta_b\cos\varphi)\gamma_A\right] \qquad \text{(VII.2.12)}$$

(we have put $\varphi = \varphi_a - \varphi_b$), the same result obtained elsewhere [2,6] through an analysis entirely based on the use of Cartesian tensors.

Let us now turn our attention to the dispersion energy. From eqs. (VII.1.15) and (VII.1.19) a convenient starting point is offered by the following expression:

$$E_{02,disp}(R,\underline{\Omega}_a,\underline{\Omega}_b) = -\frac{1}{2\pi}\sum_{L_aL_b}\sum_{L'_aL'_b}\sum_{MM'}\sum_{m_am_b}(-1)^{L_a+L'_a}$$

$$\times\left[\frac{(2L_a+2L_b+1)!\,(2L'_a+2L'_b+1)!}{(2L_a)!\,(2L_b)!\,(2L'_a)!\,(2L'_b)!}\right]^{1/2} R^{-(L_a+L_b+L'_a+L'_b+2)}$$

$$\times\begin{pmatrix} L_a & L_b & L_a+L_b \\ M & -M & 0\end{pmatrix}\begin{pmatrix} L'_a & L'_b & L'_a+L'_b \\ M' & -M' & 0\end{pmatrix}\mathbb{D}^{L_a*}_{M,m_a}(\underline{\Omega}_a)\,\mathbb{D}^{L'_a*}_{M',-m_a}(\underline{\Omega}_a)$$

$$\times\,\mathbb{D}^{L_b*}_{-M,m_a}(\underline{\Omega}_b)\,\mathbb{D}^{L'_b*}_{-M',-m_b}(\underline{\Omega}_b)\int_0^\infty d\omega\;\alpha^{L'_a,-m_a}_{L_a,m_a}(i\omega)\;\alpha^{L'_b,-m_b}_{L_b,m_b}(i\omega) \qquad \text{(VII.2.13)}$$

where we have made use of $\alpha^{L',m'}_{L,m}(i\omega) = \alpha^{L',-m}_{L,m}(i\omega)\,\delta_{m,-m'}$, an obvious generalization of eq. (VII.2.1). If we resort to the Clebsh-Gordan reduction formula for the direct product $\mathbb{D}^{L}\otimes\mathbb{D}^{L'}$, eq. (VII.2.3), we obtain

$$E_{02,disp}(R,\underline{\Omega}_a,\underline{\Omega}_b) = -2\sum_{L_aL_b}\sum_{L'_aL'_b}\sum_{MM'}\left[\frac{(2L_a+2L_b+1)!\,(2L'_a+2L'_b+1)!}{(2L_a)!\,(2L_b)!\,(2L'_a)!\,(2L'_b)!}\right]^{1/2}$$

$$\times\ R^{-(L_a+L_b+L_a'+L_b'+2)} \begin{pmatrix} L_a & L_b & L_a+L_b \\ M & -M & 0 \end{pmatrix} \begin{pmatrix} L_a' & L_b' & L_a'+L_b' \\ M' & -M' & 0 \end{pmatrix} \sum_{J_a=|L_a-L_a'|}^{L_a+L_a'}$$

$$\times \sum_{J_b=|L_b-L_b'|}^{L_b+L_b'} \begin{pmatrix} L_a & L_a' & J_a \\ M & M' & -(M+M') \end{pmatrix} \begin{pmatrix} L_b & L_b' & J_b \\ -M & -M' & M+M' \end{pmatrix} Y_{J_a}^{M+M'}(\theta_a,\varphi_a)\, Y_{J_b}^{-(M+M')}(\theta_b,\varphi_b)$$

$$\times \int_0^\infty d\omega\ \alpha_{(L_a,L_a')J_a,0}(i\omega)\ \alpha_{(L_b,L_b')J_b,0}(i\omega) \qquad \text{(VII.2.14)}$$

where, in analogy to eq. (VII.2.5),

$$\alpha_{(L,L')J,0}(i\omega) = (-1)^{L-L'}(2J+1)^{\frac{1}{2}} \sum_m \begin{pmatrix} L & L' & J \\ m & -m & 0 \end{pmatrix} \alpha_{L,m}^{L',-m}(i\omega) \qquad \text{(VII.2.15)}$$

is the j-th rank irreducible polarizability tensor prolonged in the complex plane to the imaginary axis.

From eq. (VII.2.14) we shall content ourselves to calculate the lowest order contribution to the anisotropic dispersion energy, that corresponds to $L_a=L_b=L_a'=L_b'=1$ and varies as R^{-6}. The dispersion coefficient $C_6^{disp}(\underline{\Omega}_a,\underline{\Omega}_b)$ [see eq. (VII.1.22)] is given by

$$C_6^{disp}(\underline{\Omega}_a,\underline{\Omega}_b) = 60 \sum_{M=-1}^{+1} \sum_{M'=-1}^{+1} \begin{pmatrix} 1 & 1 & 2 \\ M & -M & 0 \end{pmatrix} \begin{pmatrix} 1 & 1 & 2 \\ M' & -M' & 0 \end{pmatrix} \sum_{J_a} \sum_{J_b}$$

$$\times \begin{pmatrix} 1 & 1 & J_a \\ M & M' & -(M+M') \end{pmatrix} \begin{pmatrix} 1 & 1 & J_b \\ -M & -M' & M+M' \end{pmatrix} Y_{J_a}^{M+M'}(\theta_a,\varphi_a)\, Y_{J_b}^{-(M+M')}(\theta_b,\varphi_b)$$

$$\times \int_0^\infty d\omega\ \alpha_{(1,1)J_a,0}(i\omega)\ \alpha_{(1,1)J_b,0}(i\omega) \qquad \text{(VII.2.16)}$$

Using standard properties of 3-J coefficients, after some algebra one obtains

$$C_6^{disp}(\theta_a,\theta\ ,\varphi) = \frac{1}{4\pi}\{4\int_0^\infty d\omega\ \alpha^{(A)}_{(1,1)0,0}(i\omega)\ \alpha^{(B)}_{(1,1)0,0}(i\omega)$$

$$-\sqrt{2}[(3\cos^2\theta_a-1)\int_0^\infty d\omega\ \alpha^{(A)}_{(1,1)0,0}(i\omega)\ \alpha^{(B)}_{(1,1)2,0}(i\omega)$$

$$+ (3\cos^2\theta_b-1)\int_0^\infty d\omega\ \alpha^{(A)}_{(1,1)2,0}(i\omega)\ \alpha^{(B)}_{(1,1)0,0}(i\omega)]$$

$$+ [\frac{3}{2}\ \mathrm{sen}^2\theta_a \mathrm{sen}^2\theta_b \cos 2\varphi - 12\ \mathrm{sen}\theta_a\cos\theta_a\mathrm{sen}\theta_b\cos\theta_b\cos\varphi$$

$$+ \frac{3}{2}\ (3\cos^2\theta_a-1)(3\cos^2\theta_b-1)]\int_0^\infty d\omega\ \alpha^{(A)}_{(1,1)2,0}(i\omega)\ \alpha^{(B)}_{(1,1)2,0}(i\omega)\}$$

The irreducible tensors $\alpha_{(1,1)0,0}(i\omega)$, $\alpha_{(1,1)2,0}(i\omega)$ can be expressed in terms of Cartesian tensor components, in complete analogy with the static case; from the generalizations of eqs. (VII.2.8), (VII.2.10) and (VII.2.11), it follows

$$C_6^{disp}(\theta_a,\theta_b,\varphi) = \frac{3}{\pi}\int_0^\infty d\omega\ \bar{\alpha}_A(i\omega)\ \ \bar{\alpha}_B(i\omega) + \frac{1}{2\pi}\ [(3\cos^2\theta_a-1)$$

$$\times \int_0^\infty d\omega\ \bar{\alpha}_A(i\omega)\ \gamma_B(i\omega) + (3\cos^2\theta_b-1)\int_0^\infty d\omega\ \gamma_A(i\omega)\ \bar{\alpha}_B(i\omega)]$$

$$+ \frac{1}{4\pi}\ [\mathrm{sen}^2\theta_a\mathrm{sen}^2\theta_b\cos 2\varphi - 8\ \mathrm{sen}\theta_a\cos\theta_a\mathrm{sen}\theta_b\cos\theta_b\cos\varphi$$

$$+ (3\cos^2\theta_a-1)(3\cos^2\theta_b-1)]\int_0^\infty d\omega\ \gamma_A(i\omega)\ \gamma_B(i\omega) \qquad \text{(VII.2.17)}$$

another way of writing down a known result [4,183,184].

If we come back to eq. (VII.2.12), it is immediately seen that the lowest order contribution to the induction energy survives only if at least one of the molecules is endowed with a non-vanishing dipole moment; on the other hand, from eq. (VII.2.17) C_6^{disp} appears in any case a non-vanishing quantity.

The interaction between two <u>S-state</u> <u>atoms</u> can easily be recovered from the preceding results. Since all the permanent multipole moments vanish for a spherical atom, in this case the induction energy $E_{02,ind}$ does not play any role; as far as the dispersion energy is concerned, the lowest order contribution stems from eq. (VII.2.17) if we recognize that $\gamma = 0$, so that (omitting for simplicity the bar from $\bar{\alpha}$),

$$C_6^{disp} = \frac{3}{\pi}\int_0^\infty d\omega\ \alpha_A(i\omega)\ \alpha_B(i\omega) \qquad \text{(VII.2.18)}$$

a formally fairly simple result [2,144,184].

VII.3. Induction Energy Coefficients and Static Polarizabilities of Molecules: Two Problems in One.

In the approximation of neglecting charge overlap effects between two interacting "monomers", the multipole expansion (VI.3.2) for the interaction potential energy operator becomes adequate and the induction and dispersione energy take the simplified forms encountered in the preceding two sections [see eqs. (VII.1.8), (VII.1.14), (VII.1.15), (VII.1.21)].

Contrary to the case of the 1-st order Coulomb energy (see sec. VI. 3), whose evaluation involves only the knowledge of the permanent multipole moments of the isolate "monomers", obtaining 2-nd order energies requires in addition more complicate properties of the isolate partners, i.e. static multipole polarizabilities.

Our level of knowledge of permanent multipole moments of molecules is anything but exhaustive, the amount of reliable experimental information on multipole moments being rather scarce (a useful, although not too up-to-date, review can be found in ref. [185]). Thus, in general, we must resort to computed moments, with not negligible problems of reliability posed by the frequently insufficient quality of the electronic wavefunctions employed. A review of the methods available for the evaluation of induction energy coefficients therefore becomes largely an investigation of methods for determining static polarizabilities of atoms and molecules.

Any static multipole polarizability is, in principle, an observable property, because it characterizes the (linear) response of atoms or molecules to a suitable external electric field. Only dipole polarizabilities (and, in particular way, their isotropic components) are, at present, susceptible of experimental determination (a concise but up-to-date review can be found in ref. [186], so that our (very incomplete) knowledge of multipole polarizability data stems very much from computational efforts, which are based on a wide variety of approaches. As a general comment, it looks appropriate to point out that the problem of evaluating accurate values for polarizabilities is a hard one, the difficulties being intrinsically related to the presence in the relevant formulae of the resolvent $\hat{R}_0$ and, consequently, its adequate spectral representation.

A first, direct approach to the calculation of molecular polarizabilities relies on the availability of so-called transition oscillator strengths. For example, in the case of linear molecules (ground state of Σ symmetry), the knowledge of the two sets of oscillator strengths

$$\begin{aligned} f^{\parallel}_{k0} &= 2(E_k - E_0)\,|\langle\varphi_0|\hat{\mu}_z|\varphi_k\rangle|^2 \\ f^{\perp}_{k0} &= 2(E_k - E_0)\,|\langle\varphi_0|\hat{\mu}_x|\varphi_k\rangle|^2 \end{aligned} \qquad \text{(VII.3.1)}$$

relative to $\Sigma \to \Sigma$ and $\Sigma \to \Pi$ transitions, respectively, leads in a straightforward way to the components $\alpha^{\parallel}$ and $\alpha^{\perp}$ of the dipole polarizability, according to

$$\begin{aligned} \alpha^{\parallel} &= \sum_k{}' f^{\parallel}_{k0}(E_k - E_0)^{-2} \\ \alpha^{\perp} &= \sum_k{}' f^{\perp}_{k0}(E_k - E_0)^{-2} \end{aligned} \qquad \text{(VII.3.2)}$$

(see the preceding section). Since dipole oscillator strengths are reasonably well known for a number of optically allowed transitions of atoms and molecules, one has at disposal a technique for approaching polarizabilities and also establishing upper and lower bounds to them (the bounding property descends fundamentally from the validity of the Thomas-Reiche-Kuhn sum rule (see Appendix C), $\sum_k f^{\parallel}_{k0} = \sum_k f^{\perp}_{K0} = N$,

N being the total number of electrons in the molecule) [187-189].

The firm belief that H.F. wavefunctions are basically suitable for evaluating properties corresponding to one-electron operators has prompted some variational, perturbative and variational-perturbative methods for the computation of polarizabilities.

n the presence of an external (time-independent), homogeneous electric field $\underline{F}$, the usual electrostatic Hamiltonian operator $\hat{H}_0$ of a molecule becomes $\hat{H}_0 - \underline{F}\cdot\hat{\underline{\mu}}$, where $\hat{V} = -\underline{F}\cdot\hat{\underline{\mu}}$ represents the perturbation term corrsponding to the coupling between field and electronic motions ($\hat{\underline{\mu}} \equiv \int d\underline{r}\ \underline{r}\ \hat{\rho}^{el}(\underline{r}) = -\sum_j \underline{r}_j$ is the dipole moment operator introduced in sec. VII.2) . An approximate eigenstate in the form of a Slater determinant, built up in terms of N spinorbitals $|\chi_\gamma\rangle$, leads from the variational principle to an energy value which is a minimum as the spinorbitals satisfy the set of <u>coupled</u> H.F. equations

$$[\hat{h}_0 + \underline{F}\cdot\hat{\underline{r}} + \hat{V}^N_{HF}(\underline{F})]\,|\chi_\gamma\rangle = \varepsilon_\gamma|\chi_\gamma\rangle \qquad (\gamma=1,2,\dots,N) \qquad \text{(VII.3.3)}$$

along with the orthonormality conditions $\langle\chi_\gamma|\chi_\theta\rangle = \delta_{\gamma\theta}$.
Eqs. (VII.3.3) differ manifestly from those encountered in sec. V.4 in the presence of the external field term. It should be clear, however, that the H.F. potential $\hat{V}^N_{HF}$ depends on the external field through the spinorbitals $|\chi_\gamma\rangle$ which define the coulomb and exchange components of $\hat{V}^N_{HF}$ [see eq. (V.3.10)].

Once the previous equations have been solved self-consistently (the expansion of the spinorbitals $|\chi_\gamma\rangle$ in terms of properly chosen basis functions being the most usual procedure, so as to transform the given

---

* The perturbation term $\hat{V}=-\underline{F}\cdot\hat{\underline{\mu}}$ is very frequently explicited in terms of Cartesian components: that leads, in a natural way, to the polarizability tensor expressed in Cartesian form. The scalar product could be written also in terms of spherical vectors, as $-\sum(-1)^m\hat{Q}^m_1 F_{-m} \equiv \sum_m \hat{Q}^m_1 f_m$ (m=1,0,-1), where $\hat{\mu}_x=-2^{-\frac{1}{2}}(\hat{Q}^1_1-\hat{Q}^{-1}_1)$, $\hat{\mu}_y=2^{-\frac{1}{2}}i(\hat{Q}^1_1+\hat{Q}^{-1}_1)$, $\hat{\mu}_z=\hat{Q}^0_1$ (see footnote at pg.107). This writing provides a convenient starting point for deriving dipole polarizabilities expressed as irreducible tensors (see sects. VII.1, VII.2). A generalization to higher multipole polarizabilities follows in a simple way if we consider an external perturbing source whose effect is expressed by $\hat{V} = \sum_{\ell,m} \hat{Q}^m_\ell f_{\ell m}$. In the present context we shall use, to the largest extent, Cartesian notation.

set of integro-differential equations into an algebraic one [7,33,106]) for a suitable array of values of the field components $F_q (q \equiv x,y,z)$, a set of field-dependent spinorbitals $|\chi_\gamma(\underline{F})\rangle$ is obtained which allows to evaluate any expectation value. Thus, for example, the electric dipole moment of a molecule in the presence of an external electric field results

$$\langle\mu_q(\underline{F})\rangle = \sum_{\gamma=1}^{N} \langle\chi_\gamma(\underline{F})|\hat{\mu}_q|\chi_\gamma(\underline{F})\rangle \qquad (q\equiv x,y,z) \qquad \text{(VII.3.4)}$$

Since

$$|\chi_\gamma(\underline{F})\rangle = |\chi_\gamma^0\rangle + \sum_{q'} F_{q'}|(\partial\chi_\gamma/\partial F_{q'})_0\rangle + \cdots \qquad \text{(VII.3.5)}$$

[$|\chi_\gamma^0\rangle$ is the γ-th spinorbital in the absence of external field, while $|(\partial\chi_\gamma/\partial F_q)_0\rangle$ stands for the 1-st derivation of $|\chi_\gamma(\underline{F})\rangle$ with respect to the component F_q, evaluated at vanishing value of the field], eq. (VII.3.4) is equivalent to the expansion

$$\langle\mu_q(\underline{F})\rangle = \sum_{\gamma=1}^{N} \langle\chi_\gamma^0|\hat{\mu}_q|\chi_\gamma^0\rangle + \sum_{q'} F_{q'} \sum_{\gamma=1}^{N} [\langle\chi_\gamma^0|\hat{\mu}_q|(\partial\chi_\gamma/\partial F_{q'})_0\rangle + \text{c.c.}]$$

$$+ \ldots \equiv \langle\mu_q^0\rangle + \sum_{q'} \alpha_{qq'} F_{q'} + \frac{1}{2!} \sum_{q'q''} \beta_{qq'q''} F_{q'} F_{q''}$$

$$+ \frac{1}{3!} \sum_{q'q''q'''} \gamma_{qq'q''q'''} F_{q'} F_{q''} F_{q'''} + \ldots\ldots (q,q',\ldots \equiv x,y,z) \qquad \text{(VII.3.6)}$$

$\langle\mu_q^0\rangle$ represents the q-th component of the permanent dipole moment of the molecule, while the other contributions describe (generally non-linear) moments induced by the field. We have therefore at our disposal methods for evaluating dipole polarizability components as $\alpha_{qq'} = \lim_{q'\to 0} (\langle\mu_q(F_{q'})\rangle - \langle\mu_q^0\rangle)/F_{q'}$; hyperpolarizabilities $\beta_{qq'q''}$, $\gamma_{qq'q''q'''}$, etc. [2,190] could be obtained in a similar way.

The method now discussed, which generalizes to the case of a field dependent Hamiltonian the standard SCF procedures employed in Quantum Chemistry, is known as the finite field method [191].

Starting from the basic eqs. (VII.3.3) we are also in the position of deriving a rather extensively employed perturbative procedure, known as coupled H.F. approach [2,19,123,124] to which we have made reference in sec V.3. If we assume the external field to be very weak, so that we may limit ourselves to consider 1-st order correction to the field-free spinorbitals $|\chi_\gamma^0\rangle$ (i.e. those usually evaluated in H.F. calculations on isolate molecules), after substituting eq. (VII.3.5) into eqs. (VII.3.3) and separating off terms corresponding to different orders of magnitudo in the external field, we find

$$[\hat{h}_0 + (\hat{V}^N_{HF})_0 - \varepsilon_\gamma^0]|\chi_\gamma^0\rangle = 0$$

$$[\hat{h}_0 + (\hat{V}^N_{HF})_0 - \varepsilon_\gamma^0]|(\partial\chi_\gamma/\partial F_q)_0\rangle = [(\partial\varepsilon_\gamma/\partial F_q)_0 + \hat{r}_q - (\partial\hat{V}^N_{HF}/\partial F_q)_0]|\chi_\gamma^0\rangle \quad (\gamma = 1,2,\dots,N) \tag{VII.3.7}$$

....

along with the orthonormalization conditions $\langle\chi_\gamma^0|\chi_\theta^0\rangle = \delta_{\gamma\theta}$, $\langle\chi_\gamma^0|\,\partial\chi_\theta/\partial F_q)_0\rangle = 0$ $(\gamma,\theta = 1,2,\dots N)$.

The first set of N equations is nothing but another way of writing the equations (V.3.16) considered in sec. V.3, limitedly to the N unperturbed occupied spinorbitals $|\chi_\gamma^0\rangle$. [The H.F. operator $\hat{h}_0^{HF} = \hat{h}_0 + (\hat{V}^N_{HF})_0$ admits as solutions, in addition to the $|\chi_\gamma^0\rangle$'s, the set of unoccupied (or virtual) spinorbitals $|\chi_r^0\rangle$, $r=N+1,\dots,\infty$, with orbital energies ε_r^0; the set $\{|\chi_\gamma^0\rangle\}\oplus\{|\chi_r^0\rangle\}$ is complete, i.e. $\sum_\gamma |\chi_\gamma^0\rangle\langle\chi_\gamma^0| + \sum_r |\chi_r^0\rangle\langle\chi_r^0| \equiv \sum_i |\chi_i^0\rangle\langle\chi_i^0| = \hat{1}$].

The second set of coupled N equations (VII.3.7) allows the 1-st order corrections $|(\partial\chi_\gamma/\partial F_q)_0\rangle$ to the N spinorbitals $|\chi_\gamma^0\rangle$ to be determined. A rather popular procedure involves the expansion $|(\partial\chi_\gamma/\partial F_q)_0\rangle = \sum_r |\chi_r^0\rangle C_{r\gamma}$, by which the perturbation integro-differential equations are transformed into a set of algebraic equations for the coefficients $C_{r\gamma}$. From eq. (VII.3.6) the polarizability $\alpha_{qq'}$ can then be easily obtained. A correct expansion of $|(\partial\chi_\gamma/\partial F_q)_0>$ in terms af a basis, for example the spinorbitals $|\chi_r^0>$ above cited, would require in principle using a complete set; in that case, naturally, we would dispose of an exact spectral representation of the relevant resolvent $\hat{R}_0$ involved. Only limited basis sets of spinorbitals are actually used in any calculation, so that the problem of a truncated basis, which however warrants a practical saturation, becomes the decisive matter. The adequacy of the employed basis is obviously equally decisive when the finite field method is employed.

One should note in eq. (VII.3.7) the presence of an effective perturbation operator causing the unperturbed spinorbitals to become modified; in addition to the direct coupling with the external field, in fact, there is the effect arising from the (linear) change in the H.F. potential induced by the field $\underline{F}$. The physical content of the coupled H.F. solution with respect to an exact treatment has been clarified by using diagrammatic techniques [126,192]. The conclusions can be summarized by stating that the method at issue is characterized by three basic effects: i) a direct polarization of the various spinorbitals by the external field; ii) a self-consistency effect, which accounts for the further distortion caused by the average Coulomb and exchange field produced by the spinorbitals; iii) a correlation correction, i.e. the dynamic, "instantaneous" action of the electrons on each other. This third effect, whose adequate representation requires, according to the language of many-body theory, two- and more than two-particle excitations, is however taken into account by the method in rather incomplete way.

Simplified procedures, known as uncoupled H.F. methods, can be derived from eqs. (VII.3.7). A first, fairly rough approximation to the original coupled equations stems from the complete neglect of the change induced by the field in the H.F. potential, i.e. $(\partial\hat{V}_{HF}^N/\partial F_q)_0|\chi_\gamma^0>$ $\to 0$. The resulting N equations for the quantities $|(\partial\chi_\gamma/\partial F_q)_0>$ become a set of uncoupled equations, which is therefore easily soluble.

The most serious drawback of this uncoupled approximation lies in the unphysical presence of a "self-potential" term in $(\hat{V}_{HF}^N)_0|(\partial\chi_\gamma/\partial F_q)_0>$ which remains uncancelled after neglecting $(\partial\hat{V}_{HF}^N/\partial F_q)_0|\chi_\gamma^0>$ [125]; the poor results for the polarizabilities obtained by this method are, to a large extent, a consequence of such spurious presence.

A different uncoupled approximation can be derived from eqs. (VII.3.7) if we first cancel for each spin-orbital the unphysical "self-interaction" contained in $(\hat{V}^N_{HF})_0|(\partial\chi_\gamma/\partial F_q)_0>$ and $(\partial\hat{V}^N_{HF}/\partial F_q)_0|\chi^0_\gamma>$ and next neglect the residual term corresponding to the change in the H.F. field [125]. This new uncoupled version, which is eminently simpler than the coupled one, leads in many cases to fairly acceptable results.

A selection of computed values for the average dipole polarizability is collected in Table VII.1. Since the finite field method results here reported should be considered practically converged to their H.F. limit values [193,194], they provide a valuable test for appreciating quantitatively the reliability of that approach. The matter is of importance, in view of the possibility of extending calculations to single components of the polarizability tensor (anisotropic interactions) as well as to higher multipole polarizabilities, for both of which experiments appear nearly powerless.

If more accurate polarizability values are requested, overcoming the H.F. approximation becomes inescapable, so as to account for correlation effects to a larger extent than the coupled H.F. approach does. A particularly important role of the electron correlation is to be expected in the cases where it has an essentially non-dynamical character. Beryllium (and, more generally, alkaline earth atoms) provides a good example of such a behaviour, caused by the strong interaction between the H.F. ground state configuration $|1s^2 2s^2>$ and that doubly excited, relatively low-lying, $|1s^2 2p^2 (^1S)>$ (one also speaks frequently of near-degeneracy effect).

A double perturbation procedure of the Moller-Plesset or Epstein-Nesbet type, identical to that put forward in sec. V.4 for introducing intra-correlation corrections along perturbation lines, suggests by itself as a general approach for calculating electric and magnetic properties [107], but there have been only a few calculations which go beyond the leading term (one should recall, at this regard, that the coupled H.F. method is accurate through the 1-st order in the electron correlation). Very refined calculations of polarizabilities for

Table VII.1 - Average static dipole polarizabilities (a.u.) from H.F. theories for the ground state of some atoms and molecules (1 a.u. = $0.148176 \cdot 10^{-24}$ cm^3).

Method	Li	Be	Ne	H_2O	NH_3	CH_4	CO
Finite field (SCF)	170.3 [a]	45.63 [a]	2.368 [a]	8.50 [b]	13.35 [b]	16.00 [b]	12.40 [b]
Coupled H.F.	168÷170.3 [d-g]	42.2÷45.6 [h-k]	2.35÷2.38 [h]	7.90 [ℓ]	11.49 [m]	12.35 [n]	
Uncoupled H.F.	141.7 [p]	64.3 [p]	2.820 [p]	11.76 [q]	17.47 [q]	21.43 [r]	16.40 [r]
Experiment [s]	164±3.4		2.669	9.82	14.82	17.28	13.08

(a) Ref. [193]; (b) Ref. [194]; (d) Ref. [195]; (e) Ref. [196]; (f) Ref. [197]; (g) Ref. [198]; (h) Ref. [199]; (i) Ref. [200]; (j) Ref. [201]; (k) Ref. [202]; (ℓ) Ref. [203]; (m) Ref. [204]; (n) Ref. [205]; (p) Ref. [206]; (q) Ref. [207]; (r) Ref. [208]; (s) the experimental values are all taken from Ref. [194], except for that concerning Li, for which we have chosen a recent value reported in Ref. [186].

ground state (and also excited state) atoms have rather recently been obtained [209] by generalizing the finite field method from a SCF wavefunction to a multiconfigurational SCF wavefunction [210,211]. A quite large decrease of the polarizability, from the coupled H.F. value 45.6 a.u. to 36.5 a.u., is to be pointed out for beryllium [209]. Very promising results for dipole polarizabilities of alkaline earth atoms have been obtained very recently through an independent variation-perturbation approach which makes use of a perturbed wavefunction in the form of a very reduced expansion in terms of optimized pseudo-states [127, 212]. For beryllium, a simple "frozen core" ground state wavefunction of the type $|^1S(1s^2 2s^2)\rangle + \lambda|^1S(1s^2 2p^2)\rangle$ (variationally optimized with respect to either linear or non linear parameters), along with a perturbed wavefunction expressed as a linear combination of the configurations $|^1P(1s^2 2s2p)\rangle$ and $|^1P(1s^2 2p3d)\rangle$, after complete optimization leads to the value $\bar{\alpha} = 37.60$ a.u., in excellent agreement with the result previously quoted.

Very accurate static dipole polarizabilities for atoms and small molecules have been calculated by Meyer and coll. [193,194] on applying the finite field method to highly correlated wavefunctions arising from the Pseudo Natural Orbital-Configuration Interaction (PNO-CI) and Coupled Electron Pair Approximation (CEPA) methods [213]. A review of these approaches is out of the purposes of the present notes. A very schematic description is provided by the following excerpt from ref. [194]: "The PNO-CI yields a variational wavefunction in the form of a rapidly convergent expansion in terms of double substitution of SCF orbitals by corresponding pairs of pseudo natural orbitals,....; the PNO-CEPA wavefunction is based on the same set of configurations, but is obtained from a secular equation modified so as to account approximately for the effects of higher order substitutions". A glance to Table VII.2 shows the truly high quality of the results (one should also note that a further improvement follows from the introduction of vibrational corrections [194]).

Although our discussion has exclusively been focused on dipole polarizabilities, the computational schemes reviewed can be extended to higher multipole polarizabilities. For the average quadrupole and octopole polarizabilities of first-row atoms and ions, coupled H.F. calcu-

Table VII.2 - Average dipole polarizabilities (a.u.) from PNO-CI and PNO-CEPA methods for the ground state of some atoms and molecules.*

Method	Li	Be	Ne	H_2O	NH_3	CH_4	CO
Finite field (SCF)	170.3	45.63	2.368	8.50	13.35	16.00	12.40
PNO-CI	166.8 [a]	39.44 [a]	2.618 [a]	9.34 [b]	14.29 [b]	16.39 [b]	12.87 [b]
PNO-CEPA	164.5 [a]	37.84 [a]	2.676 [a]	9.68 [b]	14.70 [b]	16.53 [b]	13.13 [b]
Experiment	164±3.4		2.669	9.82	14.82	17.28	13.08

(a) Ref. [193]; (b) Ref. [194].

*The computed values here reported for molecules are not corrected for the vibrational motions.

lations had been carried out in the past years [195,200,214-216]. Very accurate mean quadrupole polarizability values for all neutral atoms from Li to Ne in their ground state are at present available, thanks to the efforts of Meyer and collaborators [217], who have also determined the anisotropy of the quadrupole and mixed dipole-octopole polarizabilities for the open shell atoms of the series above.

VII.4. The Evaluation of the Dispersion Energy Coefficients.

The interest in the problem at issue dates back to the thirties [46,218-224], but a really intensive research and set-up of techniques for the calculation or estimate of dispersion energy coefficients has a much shorter life, beginning from the sixties on. Our present review will be based largely on the abundant work performed in the last two decades, with some emphasis on the developments made possible by the use of the Casimir-Polder formula (see sec. VII.1), without paying attention to the historical order.

A first very valid procedure for evaluating dispersion energy coefficient, which is sometimes referred to as the summation method [144], has a generally semiempirical character, being founded on the use of oscillator strengths and resonance frequencies of atoms and/or molecules as obtained by computations and/or photoabsorption experiments. If we consider for simplicity the case of two interacting spherical atoms, from eq. (VII.1.16) or (VII.2.18) it is straightforward to show that the dipole-dipole dispersion energy coefficient C_6^{disp} can be expressed in the form

$$C_6^{disp} = \frac{3}{2} \sum_m{}' \sum_n{}' \frac{f_{mo}^A \, f_{no}^B}{E_{mo}^A \, E_{no}^B \, (E_{mo}^A + E_{no}^B)} \tag{VII.4.1}$$

where

$$f_{ko} = 2\, E_{ko} |\langle\varphi_k|\hat{\mu}_z|\varphi_o\rangle|^2 \equiv \frac{2}{3} E_{ko} |\langle\varphi_k|\hat{\underline{\mu}}|\varphi_o\rangle|^2$$

is the oscillator strength for the (optically allowed) transition $k \leftarrow o$, with resonance energy E_{ko}. More complicate expressions result in the case of interactions involving molecules; for example, the rotationally averaged dipole-dipole dispersion energy ciefficient C_6^{disp} between two linear molecules A, B is easily put in the form

$$C_6^{disp} = \frac{1}{6} \sum_n{}' \sum_m{}' \frac{4f_{no}^{A\perp} f_{mo}^{B\perp} + 2f_{no}^{A\perp} f_{mo}^{B\parallel} + 2f_{no}^{A\parallel} f_{mo}^{B\perp} + f_{no}^{A\parallel} f_{mo}^{B\parallel}}{E_{no}^{A} E_{mo}^{B} (E_{no}^{A} + E_{mo}^{B})} \qquad \text{(VII.4.2)}$$

in terms of the "longitudinal" $f^{\parallel}$ and "transversal" $f^{\perp}$ oscillator strenghts defined by eqs. (VII.3.1) (we omit the labels $\parallel$ and $\perp$ from the excitation energies E_{no}, E_{mo}).

In most cases only a limited number of the needed oscillator strenghts are known for each subsystem, but one can take advantage of the existence of sum rules (see Appendix C) to which the oscillator strenghts must obey, by neglecting values for the remaining transitions in such a way that the sum rules are satisfied by the derived set of oscillator strenghts [187,225-227].

For example, the oscillator strengths are known to satisfy the familiar and simple Thomas-Reiche-Kuhn sum rule

$$\sum_n f_{no} \equiv S(0) = N$$

N being the number of atomic electrons. At the same time, from experimental measurements of the refractive index $n(\omega)$ and Verdet constant $V(\omega) = (1/2c^2)\omega dn(\omega)/d\omega$ [228,229] we can often deduce values for some Cauchy moments $S(-2k-2)$ [eq. (C.6)]. In the normal dispersion region, in fact, the standard Lorentz-Lorenz relation between refractive index and mean dynamic dipole polarizability $\alpha(\omega)$

$$n(\omega) - 1 = 2\pi N_0 \alpha(\omega) \qquad \text{(VII.4.3)}$$

(N_0 represents the number of gaseous atoms per unit volume) can be expressed in the form [see eqs. (C.5),(C.6)]

$$n(\omega) - 1 = 2\pi N_0 \sum_{k=0}^{\infty} S(-2k-2)\,\omega^{2k} \qquad \text{(VII.4.4)}$$

The procedure consists therefore of selecting a set of oscillator strengths in such a way to reproduce S(0), S(-2), S(-4), ... and then using this set of oscillator strenghts for calculating C_6^{disp}*.

A somewhat different approach [231,232] which reduces the evaluation of the dispersion coefficients to the knowledge of moments S(-p) (p=0,-1,-2,...) can easily be appreciate if we rewrite eq. (VII.4.1) in terms of the variables

$$x_n = \frac{E_{10}^<}{E_{no}^A}, \quad y_m = \frac{E_{10}^<}{E_{mo}^B} \qquad (0 \leq x_n, y_m \leq 1)$$

where $E_{10}^< = \min(E_{10}^A, E_{10}^B)$, i.e. the smaller excitation energy between the lowest transition energies E_{10}^A, E_{10}^B with non-vanishing oscillator strengths. After some simple manipulations, eq. (VII.4.1) can be expressed in the form

$$C_6^{disp} = \frac{3}{2} (E_{10}^<)^{-3} \sum_n{}' \sum_m{}' f_{no}^A f_{mo}^B \frac{x_n^2 y_m^2}{x_n + y_m}$$

If the function $\Phi(x_n, y_m) = x_n^2 y_m^2 (x_n + y_m)^{-1}$ is approximated by a finite sum of selected powers of x_n, y_m, i.e.

$$\Phi(x_n, y_m) \simeq \sum_{pq} D_{pq} (x_n)^p (y_m)^q \qquad \text{(VII.4.5)}$$

it is straightforward to show that

$$C_6^{disp} = \frac{3}{2} \sum_{pq} (E_{10}^<)^{p+q-3} S_A(-p) S_B(-q) D_{pq} \qquad \text{(VII.4.6)}$$

where

$$S(-s) = \sum_n{}' f_{no} E_{no}^{-s}$$

---

*In the case of linear molecules, refractivity data are not sufficient for determining "longitudinal" and "transversal" Cauchy moments. The knowledge of the Rayleigh depolarization ratio for unpolarized light [229] along with the mean dipole polarizability provides expressions for polarizability components.

[see eq. (C.11)]. Different approximations of the form (VII.4.5), with constants D_{pq} chosen to make it exact, or to give a least-squares fit at some mesh of points in the range $0 \leq x_n$, $y_m \leq 1$, have been tested for the case of rare gases and atomic hydrogen: for the latter, in Table VII.3 we report a list of dispersion energy coefficients C_{2n}^{disp} (n=3,4,...,18), evaluated from generalizations of eq. (VII.4.6) to the case of higher multipoles and use of exact moments S(-1). [Terms of the form $C_{2n+1}^{disp} R^{-(2n+1)}$, n=5,6,..., arise in a natural way from 3-rd order perturbation theory. The evaluation of the corresponding coefficients has received a very limited amount of attention (see, for instance, ref. [233])].

Table VII.3. Dispersion energy coefficients (a.u.) for the interaction of two ground state hydrogen atoms [231].

n	C_{2n}^{disp}	n	C_{2n}^{disp}	n	C_{2n}^{disp}	n	C_{2n}^{disp}
3	6.500(0)*	7	0.606(7)	11	4.021(14)	15	4.553(23)
4	1.244(2)	8	3.937(8)	12	0.587(17)	16	1.181(26)
5	3.286(3)	9	3.234(10)	13	1.005(19)	17	3.458(28)
6	1.215(5)	10	3.278(12)	14	1.997(21)	18	1.13[illegible](31)

*$N(n) = N \times 10^n$

For systems other than atomic hydrogen we must expect that much less extended sets of accurately known dispersion coefficients are generally available. In some cases, however, our knowledge of dispersion coefficients is fairly wide; so, for instance, the C_6^{disp}, C_8^{disp} and C_{10}^{disp} coefficients corresponding to the interaction of S-state alkali atoms with rare gases and hydrogen atoms have been evaluated by using an approach equivalent to that just now considered in terms of the choice $\Phi(x,y) = \sum_p (-1)^p x^{2-p} y^{1+p}$. A selection of results from such reference is displayed in Table VII.4: the values refer to the interaction between ground state alkali atoms and ground state rare gases.

Table VII.4. Dispersion energy coefficients C_6^{disp}, C_8^{disp} and C_{10}^{disp} (a.u.) for interaction between alkali and rare gas atoms both in their ground state.[(a)]

	C_6^{disp}	C_8^{disp}	C_{10}^{disp}		C_6^{disp}	C_8^{disp}	C_{10}^{disp}
Li - He	2.182(1)*	1.048(3)	6.938(4)	K - He	3.407(1)	2.328(3)	2.133(5)
Li - Ne	4.256(1)	2.117(3)	1.429(5)	K - Ne	6.632(1)	4.649(3)	4.320(5)
Li - Ar	1.705(2)	8.905(3)	6.120(5)	K - Ar	2.687(2)	1.948(4)	1.828(6)
Na - He	2.410(1)	1.249(3)	8.881(4)	Cs - He	4.248(1)	3.483(3)	3.771(5)
Na - Ne	4.698(1)	2.517(3)	1.822(5)	Cs - Ne	8.266(1)	6.923(3)	7.584(5)
Na - Ar	1.887(2)	1.054(4)	7.775(5)	Cs - Ar	3.363(2)	2.898(4)	3.197(6)

*$N(n) = N \times 10^n$

(a) Proctor, T.R. and Stwalley, W.C., J. Chem. Phys. 66, 2063 (1977).

The <u>direct</u> evaluation of dispersion energy coefficients from available refractive index data, so as to avoid the necessity of intermediately obtaining Cauchy moments, is an attractive program, whose implementation is possible within the Casimir-Polder formula context. If we consider once more the case of two spherically symmetric atoms and we limit ourselves to the dipole-dipole coefficient, the working relation is given by eq. (VII.2.8)

$$C_6^{disp} = \frac{3}{\pi} \int_0^\infty d\omega\ \alpha_A(i\omega)\ \alpha_B(i\omega)$$

where $\alpha(i\omega)$ is the (average) dynamic dipole polarizability at the imaginary frequency $i\omega$. The Cauchy expansion [see App. C, eq. (C.5)] provides a convergent representation of $\alpha(\omega)$ only for frequencies $\omega < E_{10}$, the lowest resonance frequency of the right symmetry; the same is true for the polarizability at imaginary values of ω, $\alpha(i\omega)=\sum_{n=0}^{\infty}\alpha_n \cdot(-\omega^2)^n$, while the Casimir-Polder formula requires values of $\alpha(i\omega)$ on the entire positive ω semiaxis. Techniques for performing the needed analytic continuation of $\alpha(i\omega)$ along the entire semiaxis by using <u>Padé approximants</u> [230], particularly to the end of obtaining lower and upper bounds to the dispersion coefficients, will be considered in the next section; at present, we limit ourselves to assert that the dynamic polarizability $\alpha(\omega)$ can accurately be represented in the form

$$\alpha(\omega) = \sum_{m=1}^{M} \tilde{f}_m(\tilde{\omega}_m^2 - \omega^2)^{-1} \qquad \text{(VII.4.7)}$$

$\tilde{f}_m$ and $\tilde{\omega}_m$ being effective oscillator strengths and effective resonance frequencies [derived, for example, by fitting refractive index data, according to eq. (VII.4.3)]. Eq. (VII.4.7) can be extended to complex values z of the frequency and therefore evaluated, in particular, along the imaginary axis,

$$\alpha(i\omega) = \sum_{m=1}^{M} \tilde{f}_m(\tilde{\omega}_m^2 + \omega^2)^{-1} \qquad \text{(VII.4.7')}$$

If we insert eq. (VII.4.7') in the Casimir-Polder formula, the following result is easily obtained [234]

$$C_6^{disp} = \frac{3}{2} \sum_{m_A=1}^{M_A} \sum_{m_B=1}^{M_B} \frac{\tilde{f}_{m_A} \tilde{f}_{m_B}}{\tilde{\omega}_{m_A} \tilde{\omega}_{m_B} (\tilde{\omega}_{m_A} + \tilde{\omega}_{m_B})} \qquad \text{(VII.4.8)}$$

which is formally identical to a truncated version of eq. (VII.4.1). The extension of this procedure to molecules is obviously possible, provided that effective oscillator strengths and frequencies for the separate components of the dynamic polarizability are available from semiempirical or theoretical approaches [183].

After this review of a number of procedures which are essentially of a <u>semiempirical</u> character, we come to examine some <u>ab initio</u> approaches which have been used for evaluating dispersion energy coefficients.

If we look at the expression for the anisotropic dispersion energy between two arbitrary partners A,B [see eqs. (VII.1.15)], it is immediately recognized that the major problem consists of finding reliable methods for calculating the matrix elements $M^{AB}(L_a,m_a,L_b,m_b; L'_a,m'_a, L'_b,m'_b) = \langle \varphi_0^A \varphi_0^B | \hat{Q}_{L_a}^{m_a} \hat{Q}_{L_b}^{m_b} \hat{R}_0^{AB} \hat{Q}_{L'_a}^{m'_a} \hat{Q}_{L'_b}^{m'_b} | \varphi_0^A \varphi_0^B \rangle$ (for simplicity, we have omitted the superscript "el" from the electronic multipole moment operators, together with any reference to the partners involved; no confusion should arise, if we associate "a" and "b" with the partners A and B, respectively). A perturbative approach for overcoming the difficulties connected with our ignorance of exact umperturbed eigenstates is in principle possible, as shown in sec. V.4. If $|\Phi_0^A\rangle (W_0^A)$ represents the ground state eigensolution (eigenvalue) to the separable Hamiltonian operator $\hat{F}_0^A$ [eq. (V.3.3)], the lowest order correction $|\Phi_{01}^A\rangle$ to such approximate eigenstate which takes into account intra-correlation effects is given by eq. (V.3.24),

$$|\Phi_{01}^A\rangle = \hat{R}_{00}^A \; \hat{U}^A |\Phi_0^A\rangle = \hat{R}_{00}^A \; \hat{H}_0^A |\Phi_0^A\rangle$$

where $\hat{R}_{00}^A$ is the reduced resolvent,

$$\hat{R}_{00}^A = [W_0^A - \hat{F}_0^A]^{-1} [1 - |\Phi_0^A\rangle\langle\Phi_0^A|] = [W_0^A - \hat{F}_0^A]^{-1} \; \hat{P}_0^A$$

Analogous expressions can be written for $|\Phi_{01}^B\rangle$ and $\hat{R}_{00}^B$.

For simplicity, we shall go on limiting our attention to spherically symmetric partners, for which $<\Phi_0|\hat{Q}_L^m|\Phi_0> = 0$ and the appropriate expression to be considered is therefore eq. (VII.1.16).

Making use of the expansion for the resolvent $\hat{R}_0^{AB}$, eq. (V.3.26), an expansion in terms of correlation contributions to the matrix element $M^{AB} = \sum_{m_a m_b} (-1)^{m_a+m_b} M^{AB}(L_a,m_a,L_b,m_b;\ L_a,-m_a;\ L_b,-m_b)$ is obtained:

$$M^{AB} = M_0^{AB} + M_1^{AB} + \ldots\ldots \qquad \text{(VII.4.9)}$$
$$M_1^{AB} = M_{11}^{AB} + M_{12}^{AB}$$

M^{AB}, the matrix element in the approximation which neglects any correlation effect inside A and B, is clearly

$$M_0^{AB} = \sum_{m_a m_b} <\Phi_0^A\Phi_0^B|\hat{Q}_{L_a}^{m_a\dagger}\ \hat{Q}_{L_b}^{m_b\dagger}\ \hat{R}_{00}^{AB}\ \hat{Q}_{L_a}^{m_a}\ \hat{Q}_{L_b}^{m_b}|\Phi_0^A\Phi_0^B> \qquad \text{(VII.4.10)}$$

where the resolvent $\hat{R}_{00}^{AB}$ is defined as

$$\hat{R}_{00}^{AB} = [(W_0^A-\hat{F}_0^A) + (W_0^B-\hat{F}_0^B)]^{-1}\,\hat{P}_0^A\hat{P}_0^B \equiv \sum_j{}'\sum_k{}' \frac{|\Phi_j^A\Phi_k^B><\Phi_k^B\Phi_j^A|}{W_{oj}^A + W_{ok}^B} \qquad \text{(VII.4.11)}$$

(the last expression follows from a resolution of the identity in terms of eigenstates to the "unperturbed" separable Hamiltonian $\hat{F}_0 = \hat{F}_0^A + \hat{F}_0^B$).

If the matrix element M^{AB} is approximated by its lowest order term, i.e. $M^{AB} \simeq M_0^{AB}$, one deduces for the dispersion energy the same result which is obtained from the Casimir-Polder formula in conjunction with values of the dynamic polarizabilities (at imaginary frequency) as obtained from (time-dependent) uncoupled H.F. theory (see later on). In fact, from eqs. (VII.1.18) and (VII.4.11) it is a simple exercise to verify the following formal result

$$\hat{R}_{00}^{AB} = -\frac{2}{\pi}\int_0^\infty d\omega \Big[\sum_j{}' \frac{W_{jo}^A|\Phi_j^A><\Phi_j^A|}{(W_{jo}^A)^2+\omega^2}\Big]\Big[\sum_k{}' \frac{W_{ko}^B|\Phi_k^B><\Phi_k^B|}{(W_{ko}^B)^2+\omega^2}\Big] \qquad \text{(VII.4.12)}$$

$$= -\frac{2}{\pi}\int_0^\infty d\omega\,(W_0^A-\hat{F}_0^A)\,[(W_0^A-\hat{F}_0^A)^2+\omega^2]^{-1}\,\hat{P}_0^A\,(W_0^B-\hat{F}_0^B)\,[(W_0^B-\hat{F}_0^B)^2+\omega^2]^{-1}\,\hat{P}_0^B$$

Eq. (VII.4.10) therefore becomes, in complete analogy to eqs. (VII.1.19) and (VII.1.21),

$$M_o^{AB} = -\frac{1}{2\pi}\left[(2L_a+1)(2L_b+1)\right]^{\frac{1}{2}}(-1)^{L_a+L_b} \cdot \int_0^\infty d\omega\, \alpha^o_{(L_a,L_a)0,0}(i\omega)\, \alpha^o_{(L_b,L_b)0,0}(i\omega) \tag{VII.4.13}$$

where $\alpha^o_{(L,L)0,0}(i\omega) = (-1)^L(2L+1)^{-\frac{1}{2}}\sum_m \alpha^{oL,-m}_{L,m}(i\omega)(-1)^m$ [see eq. (VII.1.12)] and

$$\alpha^{oL,-m}_{L,m}(i\omega) = 2\sum_p{}' \frac{W_{po}\langle\Phi_o|\hat{Q}^m_L|\Phi_p\rangle\langle\Phi_p|\hat{Q}^{-m}_L|\Phi_o\rangle}{(W_{po})^2+\omega^2} \tag{VII.4.14}$$

is the approximate dynamic multipole polarizability at imaginary frequency as evaluated in terms of the eigensolutions to the Hartree-Fock operator $\hat{F}_o$.

M_1^{AB}, the contribution to M^{AB} that corrects M_o^{AB} for 1-st order intra-correlation effects, is much more cumbersome to be written (and evaluated!). We have

$$M_{11}^{AB} = \sum_{m_a m_b} \langle\Phi_o^A\Phi_o^B|\hat{Q}^{m_a\dagger}_{L_a}\hat{Q}^{m_b\dagger}_{L_b}\hat{R}^{AB}_{oo}(\hat{U}-W_{o1})\hat{R}^{AB}_{oo}\hat{Q}^{m_a}_{L_a}\hat{Q}^{m_b}_{L_b}|\Phi_o^A\Phi_o^B\rangle$$

$$M_{12}^{AB} = \sum_{m_a m_b}\left[\langle\Phi_o\Phi_o|(\hat{H}^A_o\hat{R}^A_{oo}+\hat{H}^B_o\hat{R}^B_{oo})\hat{Q}^{m_a\dagger}_{L_a}\hat{Q}^{m_b\dagger}_{L_b}\hat{R}^{AB}_{oo}\hat{Q}^{m_a}_{L_a}\hat{Q}^{m_b}_{L_b}|\Phi_o^A\Phi_o^B\rangle + \text{c.c.}\right] \tag{VII.4.15}$$

where $\hat{U} = \hat{U}^A + \hat{U}^B = (\hat{H}^A_o - \hat{F}^A_o) + (\hat{H}^B_o - \hat{F}^B_o)$ is the "residual interaction" operator.

Eqs. (VII.4.15) can usefully be transformed if one exploits eq. (VII.4.12) and then introduces the correction $|\Phi^{L,m}_{o1}(\omega)\rangle$ to $|\Phi_o\rangle$, formally of 1-st order, induced by a (dynamic) multipole perturbation $\hat{Q}^m_L$ and satisfying the following equation

$$[(W_0-\hat{F}_0)^2+\omega^2]\,|\Phi_{01}^{L,m}(\omega)\rangle = [\hat{Q}_L^m,\hat{F}_0]\,|\Phi_0\rangle \qquad \text{(VII.4.16)}$$

We find for M_{11}^{AB} the alternative expression

$$M_{11}^{AB} = \left(\frac{2}{\pi}\right)^2 \int_0^\infty d\omega \int_0^\infty d\omega' \cdot \sum_{m_a m_b} \langle \Phi_{01}^{L_a,m_a}(\omega)\,\Phi_{01}^{L_b,m_b}(\omega)\,|\hat{U}-W_{01}|\,\Phi_{01}^{L_a,m_a}(\omega')\,\Phi_{01}^{L_b,m_b}(\omega')\rangle \qquad \text{(VII.4.17)}$$

which involves matrix elements of the "residual interaction" operators $\hat{U}$ between single-excited configurations (as easily seen from eq. (VII. 4.16), $|\Phi_{01}^{L,m}(\omega)\rangle$ has non-vanishing amplitudes $\langle\Phi_k|\Phi_{01}^{L,m}(\omega)\rangle$ only with respect to the $|\Phi_k\rangle$'s belonging to the subset of singly-excited eigenstates to $\hat{F}_0$).

For the matrix element M_{12}^{AB} we get, in a similar way,

$$M_{12}^{AB} = \frac{1}{\pi}\int_0^\infty d\omega\, \Big\{\Big[(-1)^{L_a}(2L_a+1)^{\frac{1}{2}}\,\alpha^0_{(L_a,L_a)0,0}(i\omega) \cdot \sum_{m_b} \langle\Phi_0^B|\hat{H}_0^B\hat{R}_{00}^B\,\hat{Q}_{L_b}^{m_b\dagger}|\Phi_{01}^{L_b,m_b}(\omega)\rangle + \text{c.c.}\Big] + \Big[(-1)^{L_b}(2L_b+1)^{\frac{1}{2}}\alpha^0_{(L_b,L_b)0,0}(i\omega) \cdot \sum_{m_a} \langle\Phi_0^A|\hat{H}_0^A\hat{R}_{00}^A\,\hat{Q}_{L_a}^{m_a\dagger}|\Phi_{01}^{L_a,m_a}(\omega)\rangle + \text{c.c.}\Big]\Big\}$$

An alternative expression for M_{12}^{AB} is obtained if we introduce the correction $|\Phi_{02}^{L,m}(\omega)\rangle$ to $|\Phi_0\rangle$, formally of 2-nd order, which satisfies the equation

$$(W_0-\hat{F}_0)\,|\Phi_{02}^{L,m}(\omega)\rangle = \hat{P}_0\hat{Q}_L^{m\dagger}\,|\Phi_{01}^{L,m}(\omega)\rangle \qquad \text{(VII.4.18)}$$

M_{12}^{AB} can be rewritten as

$$M_{12}^{AB} = \frac{1}{\pi}\int_0^\infty d\omega\, \Big\{\Big[(-1)^{L_a}(2L_a+1)^{\frac{1}{2}}\alpha^0_{(L_a,L_a)0,0}(i\omega) \cdot \sum_{m_b}\langle\Phi_0^B|\hat{H}_0^B|\Phi_{02}^{L_b,m_b}(\omega)\rangle + \text{c.c.}\Big] + \Big[(-1)^{L_b}(2L_b+1)^{\frac{1}{2}}\alpha^0_{(L_b,L_b)0,0}(i\omega) \cdot \sum_{m_a}\langle\Phi_0^A|\hat{H}_0^A|\Phi_{02}^{L_a,m_a}(\omega)\rangle + \text{c.c.}\Big]\Big\} \qquad \text{(VII.4.19)}$$

Contrary to the case of M_{11}^{AB}, the matrix elements $<\Phi_0|\hat{H}_0|\Phi_{02}^{L,m}(\omega)>$ in M_{12}^{AB} have contributions only from the doubly-excited configurations appearing in the expansion of $|\Phi_{02}^{L,m}(\omega)>$; the overall correction M_1^{AB} to M_0^{AB} requires therefore the inclusion on the same foot of singly- and doubly-excited eigensolutions to $\hat{F}_0^A, \hat{F}_0^B$. In the many-body perturbation theory language [126,192] these two kinds of configurations are said to correct (at the lowest order) the simple uncoupled H.F. result for self-consistency and correlative effects, respectively (see also the preceding section).

Computational advantages can be obtained from using the expressions for M_{11}^{AB} and M_{12}^{AB} as given by eqs. (VII.4.17), (VII.4.19), because eqs. (VII.4.16) and (VII.4.18) which define the corrections $|\Phi_{01}^{L,m}(\omega)>$ and $|\Phi_{02}^{L,m}(\omega)>$ to $|\Phi_0>$ are susceptible of useful approximate solutions. So, for instance, if we minimize the variational expression

$$J_1 = <\tilde{\Phi}_{01}^{L,m}(\omega)|(W_0-\hat{F}_0)^2 + \omega^2|\tilde{\Phi}_{01}^{L,m}(\omega)> + <\tilde{\Phi}_{01}^{L,m}(\omega)|[\hat{F}_0,\hat{Q}_L^m]|\Phi_0> + <\Phi_0|[\hat{F}_0,\hat{Q}_L^m]^\dagger|\tilde{\Phi}_{01}^{L,m}(\omega)> \tag{VII.4.20}$$

with respect to reasonably flexible trial vectors $|\tilde{\Phi}_{01}^{L,m}(\omega)>$, sufficiently accurate approximations to $|\Phi_{01}^{L,m}(\omega)>$ can be evaluated [144,235].

A systematic approach to the calculation of the dispersion energy coefficients along the lines now recalled (<u>double perturbation theory</u>), although possible, has been applied in a very limited number of cases [121,122]. More frequently, self-consistency and correlative corrections through the 1-st order have been introduced by using <u>time-dependent coupled H.F. theory</u> [2,19,144,236-239], an approximation which generalizes to the case of time-dependent perturbations the approach discussed in the preceding section. Since the derivation of the working equations is a rather involved procedure, we have confined the details to the Appendix D, being here satisfied with a sketchy review of the approach.

The method under investigation essentially resolves to approximate the state of a N electron system, perturbed by a weak time-dependent external field, in the form of a Slater determinant $|\Phi(t)> = (N!)^{-\frac{1}{2}}$

$\det[|\chi_1(1;t)\rangle\,|\chi_2(2;t)\rangle\ldots|\chi_N(N;t)\rangle]$ built up in terms of field-dependent spinorbitals $|\chi_\alpha(t)\rangle$, the approximation being supposed valid at any time. If each field-dependent spinorbital is represented in terms of the unperturbed spinorbitals $|\chi_j^0\rangle$, solutions to the H.F. equations in the absence of external field, i.e. $|\chi_\alpha(t)\rangle = |\chi_\alpha^0\rangle + \sum_r |\chi_r^0\rangle C_{r\alpha}(t)$ [where r,s (α,β) label unoccupied (occupied) spinorbitals in the H.F. ground state $|\Phi_0\rangle = (N!)^{-\frac{1}{2}}\det(|\chi_1^0(1)\rangle\ldots|\chi_N^0(N)\rangle)$ of the unperturbed system], it is rather easy to verify that the simple determinant $|\Phi(t)\rangle$ is equivalent to an expansion in terms of singly-excited $|\Phi_\alpha^r\rangle$, doubly-excited $|\Phi_{\alpha\beta}^{rs}\rangle,\ldots$ configurations,

$$|\Phi(t)\rangle = |\Phi_0\rangle + \sum_{r\alpha} |\Phi_\alpha^r\rangle\, C_{r\alpha}(t) + \frac{1}{2}\sum_{r\alpha}\sum_{s\beta} |\Phi_{\alpha\beta}^{rs}\rangle\, C_{r\alpha}(t)\,C_{s\beta}(t) + \ldots. \quad \text{(VII.4.21)}$$

The coupled H.F. procedure can therefore be founded on the use of the approximate perturbed state given in eq. (VII.4.21), the unknown quantities being the time-dependent coefficients $C_{r\alpha}(t)$.

If we assume a time-dependent external perturbation of monochromatic type

$$\hat{V}(t) = \hat{v}\, e^{-i\omega t} + \hat{v}^{\dagger}\, e^{i\omega t} \quad \text{(VII.4.22)}$$

$C_{r\alpha}(t) = Y_{r\alpha}(\omega)\, e^{-i\omega t} + Z^*_{r\alpha}(\omega)\, e^{i\omega t}$ suggests itself as a natural ansatz , which reduces the determination of the approximate solution $|\Phi(t)\rangle$ to that of the frequency-dependent coefficients $Y_{r\alpha}(\omega)$, $Z_{r\alpha}(\omega)$. As shown in Appendix D, the sought solution can be cast into the compact form

$$\begin{aligned}(\mathbb{A} - \omega\mathbb{1})|\mathbb{Y}\rangle + \mathbb{B}\,|\mathbb{Z}\rangle &= -\,|\mathbb{v}\rangle\\ (\mathbb{A} + \omega\mathbb{1})|\mathbb{Z}\rangle + \mathbb{B}\,|\mathbb{Y}\rangle &= -\,|\mathbb{v}\rangle\end{aligned} \quad \text{(VII.4.23)}$$

where $|\mathbb{v}\rangle$ is a column supervector of elements $\langle\Phi_\alpha^r|\hat{v}|\Phi_0\rangle$, $|\mathbb{Y}\rangle$ and $|\mathbb{Z}\rangle$ column supervectors of elements $Y_{r\alpha}(\omega)$ and $Z_{r\alpha}(\omega)$, while $\mathbb{A}$ and $\mathbb{B}$ are supermatrices of elements $\mathbb{A}_{r\alpha,s\beta} = \langle\Phi_\alpha^r|\hat{H}_0 - E_{HF}|\Phi_\beta^s\rangle$, $\mathbb{B}_{r\alpha,s\beta} = \langle\Phi_{\alpha\beta}^{rs}|\hat{H}_0|\Phi_0\rangle$

(E_{HF} is the unperturbed H.F. energy of the system).

Approximate expectation values for physical properties at any time t can now be obtained; so, for instance, one finds for the dipole moment expectation value $\langle\hat{\mu}_z\rangle_t$ in the presence of a homogeneous electric field F_z,

$$\langle\hat{\mu}_z\rangle_t = \langle\hat{\mu}_z\rangle_o + 2\,F_z\,(e^{-i\omega t}+e^{i\omega t}) \cdot \langle \boldsymbol{\mu}_z | [(\mathbb{A}-\mathbb{B})(\mathbb{A}+\mathbb{B})-\omega^2\mathbb{1}]^{-1}(\mathbb{A}-\mathbb{B})|\boldsymbol{\mu}_z\rangle \qquad \text{(VII.4.24)}$$

where $\langle\hat{\mu}_z\rangle_o \equiv \langle\Phi_o|\hat{\mu}_z|\Phi_o\rangle$ is the permanent dipole moment of the system and $|\boldsymbol{\mu}_z\rangle$ the column supervector of elements $\langle\Phi_\alpha^r|\hat{\mu}_z|\Phi_o\rangle$. Eq. (VII.4.24) allows one to identify the zz-component of the dynamic dipole polarizability tensor $\alpha_{zz}(\omega)$,

$$\alpha_{zz}(\omega) = 2\,\langle\boldsymbol{\mu}_z|[(\mathbb{A}-\mathbb{B})(\mathbb{A}+\mathbb{B})-\omega^2\mathbb{1}]^{-1}(\mathbb{A}-\mathbb{B})|\boldsymbol{\mu}_z\rangle \qquad \text{(VII.4.25)}$$

An interesting elaboration of eq. (VII.4.25) is in terms of the eigenvalues and eigenvectors to the (<u>not self-adjoint</u>) matrices $(\mathbb{A}-\mathbb{B})(\mathbb{A}+\mathbb{B})$ and $(\mathbb{A}+\mathbb{B})(\mathbb{A}-\mathbb{B})$,

$$\begin{aligned}(\mathbb{A}-\mathbb{B})(\mathbb{A}+\mathbb{B})|\mathbb{C}_n\rangle &= \omega_n^2\,|\mathbb{C}_n\rangle \\ (\mathbb{A}+\mathbb{B})(\mathbb{A}-\mathbb{B})|\mathbb{D}_n\rangle &= \omega_n^2\,|\mathbb{D}_n\rangle\end{aligned} \qquad \text{(VII.4.26)}$$

(the equality of the eigenvalues pertaining to these different equations is easily demonstrated). We can rewrite eq. (VII.4.25) in the form

$$\alpha_{zz}(\omega) = 2\sum_n \frac{\langle\boldsymbol{\mu}_z|\mathbb{C}_n\rangle\langle\mathbb{D}_n|(\mathbb{A}-\mathbb{B})|\boldsymbol{\mu}_z\rangle}{\omega_n^2-\omega^2} \qquad \text{(VII.4.27)}$$

which suggests to interpretate the eigenvalues ω_n^2 to $(\mathbb{A}-\mathbb{B})(\mathbb{A}+\mathbb{B})$ as (square) <u>excitation energies</u> of the unperturbed system.

Before returning to our main subject, we would like to point out that the coupled H.F. procedure is a many-electron theory which includes, at the lowest order, the self-consistency and correlation effects more than once recalled. One should also mention that there exist more general approaches to the very difficult problem of accounting for the dynamical effects involved in the excitations of many-body systems. These approaches make use of various formalisms (Green's functions, polarization propagators, equations-of-motion methods) [240-242]: it is of interest to know that the physical content of the coupled H.F. theory is the same as in these more general many-body procedures, when they are solved at a (low) order of approximation (the so-called Random Phase Approximation).

The coupled H.F. theory leads to an expression for the dynamic polarizability $\alpha_{zz}(\omega)$ with the same form as eq. (VII.4.7.), so that its extension to complex values of the frequency ω is quite natural. From $\alpha_{zz}(i\omega) = 2 \sum_n (\omega_n^2+\omega^2)^{-1} < \mathbb{\mu}_z | \mathbb{C}_n >< \mathbb{D}_n | (\mathbb{A}-\mathbb{B}) | \mathbb{\mu}_z >$, it is then an easy task from the Casimir-Polder formula to attain to a result for C_6^{disp} having the same form as eq. (VII.4.8). It should also be clear that higher dispersion coefficients C_8^{disp}, C_{10}^{disp},... can be evaluated by the same approach now applied to C_6^{disp}, through the introduction in the formalism of suitable external multipole fields. We point out in the present context that dynamic polarizabilities extended at imaginary values of the frequency are <u>authomatically</u> generated by using external fields with a time dependence of the form $e^{-\omega t}$ [see eq. (VII.4.22)]: such remark [243] has been usefully implemented in a few cases [121, 122,244,245]. Actual calculations are performed in any case in terms of truncated sets of spinorbitals so that their number and quality become crucial elements for achieving convergent results. H.F. unoccupied (virtual) spinorbitals, solutions to the standard H.F. equations in a $\hat{V}^N$ potential, display a not ideal behaviour from this standpoint (see also sec. V.4) contrary to the spinorbitals obtained in a $\hat{V}^{N-1}$ potential.

A number of dispersion energy coefficients C_{2n}^{disp} are collected in Table VII.5. In any case they correspond to <u>isotropic</u>, i.e. rotationally averaged, quantities. Anisotropic coefficients, although available for some of the cases there reported, have not been included in order not to encumber the Table. The interactions considered have been chosen

Table VII.5. Dispersion energy coefficients C_{2n}^{disp} (a.u.) for a number of typical interactions, according to various computational procedures.

	He - He			Li-Li	Be - Be		Ne - Ne		H_2 - H_2		H_2O - H_2O		C_2H_4 - C_2H_4	
	C_6^{disp}	C_8^{disp}	C_{10}^{disp}	C_6^{disp}	C_6^{disp}	C_8^{disp}	C_6^{disp}	C_8^{disp}	C_6^{disp}	C_8^{disp}	C_6^{disp}	C_8^{disp}	C_6^{disp}	C_8^{disp}
uncoupled H.F. theory	1.66[a,b] 1.664[c]	14.65[c]	182.37[c]				7.727[q]	87.03[q]						
coupled H.F. theory	1.376[d] 1.370[e]	12.98[e]	163.58[e]	1479[m]	312.8[e]	14320[e]	5.44[r]		11.50[r']		372[v]			
double pert. theory	1.335[c]	13.00[c]	168.42[c]				5.308[q]	61.18[q]						
variation perturbation				1453[n]	221.04[o]				13.12[s]				398.7[y]	16890[y]
strongly correlated calculations	1.457[f] 1.456[g]	13.88[f] 13.90[g]	176.93[f] 175.4[g]		213.5[f] 213.47[p]	9088[f]			12.14[g]	215.2[g]				
oscillator model	0.83[h]	5.42[h]	43.3[h]						11.6[h]	182[h]	36.5[h]	333[h]	321[h]	4660[h]
semiempirical	1.47[i]			1390[i]			6.3[i]		13[i] 12.4[t]	116[u]	45.4[w]		303[z]	
accurate	1.458[j] 1.4605[k] 1.460[ℓ]													

(a) Ref. [246]; (b) Ref. [144]; (c) Ref. [121]; (d) Ref. [247]; (e) Ref. [245]; (f) Ref. [248]; (g) Ref. [249]; (h) Ref. [250]; (i) Semiempirical estimate recommended by A. Dalgarno, ref. [2]; (j) Ref. [251]; (k) Ref. [252]; (ℓ) Ref. [253]; (m) Ref. [254]; (n) Ref. [127]; (o) Ref. [212]; (p) Ref. [255]; (q) Ref. [122]; (r) Ref. [256]; (r') Ref. [257]; (s) Ref. [258]; (t) Ref. [183]; (u) Ref. [259]; (v) Ref. [260]; (w) Ref. [261]; (y) Ref. [159]; (z) Ref. [262].

somewhat arbitrarily (the choice is not so large, however), with the primary intent of providing some matter for comparisons between different computational approaches.

The uncoupled H.F. theory results reported correspond to the modified approximation touched upon in sec. VII.3, with unphysical "self-interaction" terms omitted [262]. The good agreement between coupled H.F. and double perturbation theory at 1-st order for the interactions He-He and Ne-Ne is not unexpected, and reflects the similarity of physical content of these two approximations. Although putting forward quantitative assessments about the reliability of the various computational methods is a hard engagement, an error of 5÷10% affecting coupled H.F. predictions of good quality (i.e., essentially free from basis set extension effects) with respect to admittedly accurate results appears a reasonable estimate, except for systems characterized by low-lying doubly excited states, strongly interacting with the H.F. ground state, in which case a strong deterioration of the coupled H.F. result occurs (see, for example, the case Be-Be). This behaviour is entirely analogous to that put in evidence in sec. VII.3 in connection with the evaluation of static polarizabilities.

Overcoming the inadequacy of the H.F. model requires more sophisticated procedures, by which to take into account intra-correlation effects. We point out in this context the rather limited amount of existing calculations, most of which carried out within a <u>variation-perturbation</u> scheme. It is not difficult, in fact, to persuade oneself that the (isotropic) dispersion coefficients C_{2k}^{disp} can also be calculated by minimizing the Hylleraas-type functionals $F_{2k}[\tilde{\Phi}_k] = \langle\tilde{\Phi}_k|(\hat{H}_0^A+\hat{H}_0^B) - (E_0^A+E_0^B)|\tilde{\Phi}_k\rangle + 2\,\mathrm{Re}\langle\tilde{\Phi}_k|\hat{V}_k|\varphi_0^A\varphi_0^B\rangle$ $(k = 3,4,\ldots)$ with respect to suitable trial vectors $|\tilde{\Phi}_k\rangle$. $R^{-3}\hat{V}_3$, $R^{-4}\hat{V}_4,\ldots$ are operators corresponding to the dipole-dipole, dipole-quadrupole,... interaction, respectively [see eq. (VII.1.16)]. As known, rigorous bounds to the true C_{2k}^{disp} dispersion coefficients result from such minimization procedures only <u>if</u> $|\varphi_0^A\varphi_0^B\rangle \equiv |\varphi_0^A\rangle|\varphi_0^B\rangle$ is an <u>exact</u> ground state eigenstate to $(\hat{H}_0^A+\hat{H}_0^B)$. Now, although exact unperturbed eigenstates are generally unavailable, correlated approximations to them can be built up; the same is true for the trial state vectors $|\tilde{\Phi}_n\rangle$, which can be chosen in the form of more or less extended multiconfiguration functions [248,249]. The values of the dispersion energy coefficients for the interactions He-He, Be-Be and H_2-H_2 reported in Table VII.5 under the heading "strongly correlated calculations" are to be considered very accurate, so that their right place could have been in the last row of the table ("accurate"). The values reported for the interaction between two ethylene molecules are presumably rather accurate and deserve consequently a particular mention, in view of the noticeable complexity of the system [159]. The same variation-perturbation approach as above has successfully been applied to simple interactions [127,212,258] by using rather simple trial wavefunctions, in the form of linear combination of a few pseudo-state configurations (even a single pseudo-state configuration) to be optimized with respect to linear and non-linear parameters (orbital exponents).

Before ending this section, we like to outline an interesting approach suggested in very recent years [250,264] for estimating dispersion energy coefficients, whose value lies mainly in the relative ease of application to molecules of sizeable dimensions. The basis of the method amounts essentially to replacing the molecular unperturbed Hamiltonian $\hat{H}_0$ by a new one $\hat{H}_{00}$, the sum of one-electron <u>harmonic oscillator</u> Hamiltonians, so that an effective oscillator model is obtained [4]. Although corrections for attenuating the crudity of the model are in principle possible in terms of a perturbative procedure involving the "perturbation" $(\hat{H}_0-\hat{H}_{00})$, they are actually ignored. The ground state wavefunction is then a Slater determinant of Gaussian spherical spinorbitals $|\chi_i\rangle = |\psi_i\rangle|\pm\frac{1}{2}\rangle$, where the orbitals $\psi_\alpha(\underline{r}) = (\omega_\alpha/\pi)^{3/4} \cdot \exp[-\frac{1}{2}\omega_\alpha|\underline{r}-\underline{R}_\alpha|^2]$ have exponents ω_α and centres $\underline{R}_\alpha$ to be determined variationally.

Now, classical and quantum theory lead to <u>identical</u> results for many properties of harmonic oscillator systems, so that it is not surprising that the dynamic dipole polarizability of the model under investigation has the following "classical" form

$$\alpha(\omega) = 2 \sum_{\alpha=1}^{N} (\omega_\alpha^2 - \omega^2)^{-1} \tag{VII.4.28}$$

where 2N is the number of electrons in the molecule. Eq. (VII.4.28) can be prolonged to complex values of the frequency ω in a straightforward way and the resulting $\alpha(i\omega)$ used in the Casimir-Polder formula. The dispersion energy coefficients reported in Table VII.5 in correspondence with the heading "oscillator model" refer to values obtained by the method now reviewed. As a comment, we limit ourselves to point out the unsatisfactory prediction of the dispersion coefficients from this model in the case of the interaction He-He, and the considerable improvement of agreement when passing to more complex systems (the result for the dipole-quadrupole dispersion coefficient C_8^{disp} in the case of the interaction C_2H_4-C_2H_4, however, deviates strongly from a different estimate [159]).

VII.5. Determination of Upper and Lower Bounds to Dispersion Energy Coefficients.

In view of the difficulties involved in calculating dispersion energy coefficients, we have seen how one is generally forced to introduce various approximations, which then affect the results with rather hardly assessable uncertainties. For this simple reason, one should be extremely satisfied if (possibly) uncomplicated procedures could be put forward, for determining bounds to the quantities of interest.

A formal analysis which leads to upper and lower bounds for 2-nd order energies in perturbation theory has been reviewed in sec. III.6, and applications of such approach will be examined later on. We start, however, by considering some alternative bounding procedures, which are based on the use of Padé approximants and moment theory, because of their noticeable importance.

It has been shown, in Appendix C, that the Cauchy dispersion expansion for the dynamic polarizability on the imaginary frequency axis (see eq. C.5),

$$\alpha(i\omega) = \sum_{n=0}^{\infty} \alpha_n (-\omega^2)^n \qquad \text{(VII.5.1)}$$

which converges only within the interval $0 \le \omega \le E_{10}$ (i.e. below the first reasonance frequency), is a series of Stieltjes in the variable $z = \omega^2$ [228-230,247]. This peculiarity has an important bearing on the successive problem of evaluating the Casimir-Polder integral, because techniques are available which allow to perform an analytic continuation of eq. (VII.5.1) outside the radius of convergence E_{10}. The Padé approximants offer a powerful way for accomplishing accurately such a program, providing in addition lower and upper bounds to the correct continuation along the positive imaginary semiaxis.

Given the series of Stieltjes (VII.5.1), the $[n,m]_\alpha$ Padé approximant to the polarizability α is defined as the following ratio

$$[n,m]_\alpha = P_m(\omega) \,/\, Q_n(\omega) \tag{VII.5.2}$$

where

$$\begin{aligned} P_m(\omega) &= a_0 + \sum_{i=1}^{m} a_i(-\omega^2)^i \\ Q_n(\omega) &= 1 + \sum_{i=1}^{n} b_i(-\omega^2)^i \end{aligned} \tag{VII.5.3}$$

are polynomials in the variable $(-\omega^2)$, respectively of order m and n. The unknown coefficients a_i, b_i appearing in the polynomials $P_m(\omega)$ and $Q_n(\omega)$ are determined by requiring that the power series expansion of eq. (VII.5.2), in the variable $(-\omega^2)$, equal eq. (VII.5.1) term by term to order $n+m$; in an equivalent way, one sets

$$P_m(\omega) - Q_n(\omega)\Big[\sum_{k=0}^{n+m} \alpha_k(-\omega^2)^k\Big] = 0 \tag{VII.5.4}$$

term by term, to order $n+m$.

The Padé approximants $[n,n]_\alpha$ and $[n,n-1]_\alpha$ are particularly important; it can be shown, in fact, that in addition to performing a continuation of eq. (VII.5.1), they provide us with lower and upper bounds to the correct continuation along the imaginary semiaxis [265-267]

$$[n,n-1]_\alpha \le \alpha(i\omega) \le [n,n]_\alpha\,, \quad 0 \le \omega < \infty \tag{VII.5.5}$$

If we consider the Padé approximants $[n,n+k]_\alpha$ (k=0,-1), the b_j coefficients for both $[n,n-1]_\alpha$ and $[n,n]_\alpha$ are obtained from the linear equations [228,230,257]

$$\sum_{j=1}^{n} A_{ij}\, b_j = -\alpha_{n+k+i} \;,\; i=1,2,\ldots,n \;,\; k=0,-1$$

$$A_{ij} = \alpha_{n+k+i-j} \qquad \text{(VII.5.6)}$$

in terms of known Cauchy moments α_p. The existence of nontrivial solution to eqs. (VII.5.6) is assured by the fact that the matrix of elements A_{ij} is surely nonsingular [see eq. (C.9)]. For the a_j coefficients, one finds, in a similar way,

$$\sum_{j=1}^{i} \alpha_{i-j}\, b_j = a_i \quad , \quad i=1,2,\ldots,n+k \;,\; k=0,-1 \qquad \text{(VII.5.6')}$$

From the Casimir-Polder formula [eq. (VII.2.18)], we then obtain the following bounds to the dipole-dipole dispersion energy coefficient C_6^{disp},

$$C_6^{disp} \le \frac{3}{\pi}\int_0^\infty d\omega\, [n,n]_{\alpha_A} [n,n]_{\alpha_B}$$

$$n=1,2,\ldots \qquad \text{(VII.5.7)}$$

$$C_6^{disp} \ge \frac{3}{\pi}\int_0^\infty d\omega\, [n,n-1]_{\alpha_A} [n,n-1]_{\alpha_B}$$

the Padé approximants involved being obviously approximations to the dipole polarizabilities $\alpha_A(i\omega)$, $\alpha_B(i\omega)$.

The lower bound to C_6^{disp} can be expressed in a simple form through the following result (see also sec. VII.4),

$$[n,n-1]_\alpha = \sum_{j=1}^{n} (\omega_j^2+\omega^2)^{-1} f_j \le \alpha(i\omega) \qquad \text{(VII.5.8)}$$

which stems from a partial fraction reduction of $[n,n-1]_\alpha$ [228,230,257, 265]. In eq. (VII.5.8) f_j and ω_j represent "effective" oscillator strengths and transition frequencies, respectively, so that the Padé approximant $[n,n-1]_\alpha$ can be regarded as a way for generating a fictitious, finite spectrum to approximate the true (infinite) one of the molecule. It should be noted that the quantities f_j and ω_j are upper bounds respectively to the first n oscillator strengths f_{no} and transition frequencies E_{no} [230,265]. If eq. (VII.5.8) for $[n,n-1]_\alpha$ is substituted in eq. (VII.5.7), it is straightforward to get the result

$$\frac{3}{2}\sum_{i=1}^{n_A}\sum_{j=1}^{n_B}\frac{f_i^A f_j^B}{\omega_i^A\omega_j^B(\omega_i^A+\omega_j^B)} \leq C_6^{disp} \qquad \text{(VII.5.9)}$$

[see eq. (VII.4.8)]. The equality in this equation holds only in the limit $(n_A,n_B)\to\infty$; in spite of this fact, from a fictitious spectrum, which is usually a poor approximation to the true atomic (or molecular) one, we get accurate dispersion energy coefficients thanks to the fast numerical convergence of eq. (VII.5.9).

Contrary to $[n,n-1]_\alpha$, which is seen to satisfy the limit behaviour $\lim_{\omega\to\infty}[n,n-1]_\alpha = 0$ (the same which $\alpha(i\omega)$ is endowed with), the approximant $[n,n]_\alpha$ approaches the limit value (a_n/b_n) as $\omega\to\infty$ and the upper bounds resulting from eq. (VII.5.7) are divergent, thus providing a hardly useful result. This difficulty can be remedied by introducing the function

$$\beta(i\omega) = N-\omega^2\alpha(i\omega) = \sum_k \frac{E_{ko}f_{ko}}{E_{ko}^2+\omega^2} \quad ; \qquad \text{(VII.5.10)}$$

here the last equality involving transition frequencies E_{ko} and oscillator strengths f_{ko} stems from using the Thomas-Reiche-Kuhn sum rule $\sum_k f_{ko} = N$, with N number of electrons in the molecule (see Appendix C).

The character of series of Stieltjes for the expansion of $\beta(i\omega)$ in powers of ω^2 can be demonstrated by resorting to the same procedure as in Appendix C for $\alpha(i\omega)$; from this behaviour, in analogy with eq. (VII.5.5) it follows

$$[n,n-1]_\beta \leq \beta(i\omega) = N-\omega^2\alpha(i\omega) \qquad \text{(VII.5.11)}$$

i.e.

$$\frac{N-[n,n-1]_\beta}{\omega^2} \geq \alpha(i\omega) \qquad \text{(VII.5.11')}$$

$[n,n-1]_\beta$ being the Padé approximant to $\beta(i\omega)$. A new upper bound to $\alpha(i\omega)$ is thus obtained, which can be cast into the form [see eq. (VII. 5.8)]

$$\frac{N - [n,n-1]_\beta}{\omega^2} = \sum_j^n \frac{\bar{f}_j}{\bar{\omega}_j^2 + \bar{\omega}^2} \geq \alpha(i\omega) \qquad \text{(VII.5.12)}$$

and does not suffer from the drawback of $[n,n]_\alpha$. If we make use of eq. (VII.5.12) in the Casimir-Polder formula, we obtain finally

$$\frac{3}{2} \sum_{j=1}^{n_A} \sum_{k=1}^{n_B} \frac{\bar{f}_j^A \bar{f}_k^B}{\bar{\omega}_j^A \bar{\omega}_k^B (\bar{\omega}_j^A + \bar{\omega}_k^B)} \geq C_6^{disp} \qquad \text{(VII.5.12)}$$

a useful upper bound to C_6^{disp}.

The use of Padé approximant techniques for establishing bounds to dispersion energy coefficients can obviously be generalized to higher multipole coefficients and to the case of anisotropic interactions as well [230,184].

Before ending this short review on the important role played by the Padé approximants in the context of the dispersion energy coefficient evaluation, it is of interest to inspect the expressions for the bounds arising from the Padé approximants $[1,0]_\alpha$, $[1,0]_\beta$. From eq. (VII.5.4), we find after a very simple algebra

$$[1,0]_\alpha = \frac{\alpha_0}{1 + \frac{\alpha_1}{\alpha_0}\omega^2} \leq \alpha(i\omega)$$

$$\frac{N - [1,0]_\beta}{\omega^2} = \frac{N}{\frac{N}{\alpha_0} + \omega^2} \geq \alpha(i\omega)$$

α_0 and α_1 being the first two moments in the Cauchy expansion for the polarizability $\alpha(\omega)$ [see eq. (C.5)]. From the Casimir-Polder formula, we get the following bounds [230,257,268]

$$\frac{3}{2}\alpha_0^A \alpha_0^B \frac{\omega_A \omega_B}{\omega_A + \omega_B} \leq (C_6^{disp})_{AB} \qquad \text{(VII.5.13)}$$

$$\frac{3}{2}\,\alpha_0^A\,\alpha_0^B\,\frac{\bar{\omega}_A\,\bar{\omega}_B}{\bar{\omega}_A+\bar{\omega}_B} \geq (C_6^{disp})_{AB} \qquad \text{(VII.5.14)}$$

where we have put

$$\omega = (\alpha_0/\alpha_1)^{\frac{1}{2}}\ ,\ \bar{\omega} = (N/\alpha_0)^{\frac{1}{2}} \qquad \text{(VII.5.15)}$$

and A, B label different interacting partners. Eq. (VII.5.13) has the same form as the well known London formula [4,219], originally obtained by using the Unsöld closure approximation. Eq. (VII.5.14) is identical to the Slater-Kirkwood formula [4,173,220,270], first derived by using a variational approach.

We pass now to consider the moment theory method [271-278], limiting ourselves for simplicity to the case of the dipole-dipole dispersion coefficient C_6^{disp} between two identical spherical atoms. If we make use of the results in Appendix C, the dynamic dipole polarizability at imaginary frequency $\alpha(i\omega)$ can be expressed in the following integral form,

$$\alpha(i\omega) = \frac{1}{2}\int_0^{E_{10}^{-2}} du\ \left(\frac{df}{d\varepsilon}\right)\frac{u^{-\frac{1}{2}}}{1+\omega^2 u}$$

$$= \int_0^{E_{10}^{-2}} du\,\frac{u^2}{1+\omega^2 u}\ \frac{1}{2}\,u^{-5/2}\ \left(\frac{df}{d\varepsilon}\right)$$

This result is easily obtained on introducing the oscillator strength distribution $(df/d\varepsilon)$ [see eq. (C.2)] and the variable $u=\varepsilon^{-2}=(E-E_0)^{-2}$ ($E-E_0$ is the excitation energy from the ground state and $E_{10}=E_1-E_0$ the lowest optically allowed transition frequency of the atom). If we then introduce the related (nonnegative) distribution function

$$F(u) = \frac{1}{2}\,u^{-5/2}\ \left(\frac{df}{d\varepsilon}\right)$$

$\alpha(i\omega)$ can be cast into the final form

$$\alpha(i\omega) = \int_0^{E_{10}^{-2}} du\ \frac{u^2}{1+\omega^2 u}\ F(u) \qquad \text{(VII.5.16)}$$

so that the evaluation of the polarizability $\alpha(i\omega)$ is essentially controlled by our knowledge of F(u). As a rule, the distribution function F(u) is unknown, but we dispose of information about it through a (usually limited) number of moments μ_k ($k = 1,2,\ldots N_m$) generated by suitable moment-generating functions. The moments to our disposal consist most frequently of a set of values of the dynamic polarizability at different frequencies,

$$\alpha(\omega_i) = \int_0^{E_{10}^{-2}} du \frac{u^2}{1-\omega_i^2 u} F(u) \qquad \text{(VII.5.17)}$$

in conjunction with the Thomas-Reiche-Kuhn sum rule [see also eq. (C.13)],

$$N = \int_0^{E_{10}^{-2}} du\, u\, F(u) \qquad \text{(VII.5.18)}$$

(N denotes the number of electrons in the atom). To these, one can add the additional moment

$$\frac{4}{3} \pi N \langle\varphi_0| \sum_{j=1}^{N} \delta(\underline{r}_j) |\varphi_0\rangle = \int_0^{E_{10}^{-2}} du\, F(u) \qquad \text{(VII.5.19)}$$

$\delta(\underline{r})$ is the Dirac delta function at the point $\underline{r}$ far from the nucleus, so that the left-hand side quantity is strictly related to the value of the electronic charge density at the nuclear position in the ground state (all these moments have already been introduced in Appendix C in terms of a distribution function other than F(u)). Unlike the moments expressed by eqs. (VII.5.17) and (VII.5.18), the moment defined by eq. (VII.5.19) does not seem to be susceptible of experimental measurement. Approximate estimates for such quantity, however, do not appear to influence critically the results we are about to deduce.

Moment theory provides us with a powerful way for establishing bounds to the unknown moments $\alpha(i\omega_j)$, eq. (VII.5.16), starting from a knowledge of known moments, for example those defined by eqs. (VII.5.17) - (VII.5.19). Without going into details, we limit ourselves to present some important results of moment theory, assuming that we dispose of N_m moments μ_k of the distribution function F(u) [278]

$$\mu_k = \int_a^b du\, f_k(u)\, F(u) \quad , \ (k = 1,2,\ldots N_m) \qquad \text{(VII.5.20)}$$

$\{f_k(u)\}$ being the set of N_m moment-generating functions. To start with, we define the principal representation of the moments μ_k as the distribution function of the form

$$F_p(u) = \sum_j W_j\, \delta(u - u_j) \qquad \text{(VII.5.21)}$$

with weights W_j and abscissae u_j which allow to reproduce the N_m known moments:

$$\begin{aligned} \mu_k &= \int_a^b du\, f_k(u)\, F_p(u) \\ &= \sum_j W_j\, f_k(u_j) \quad , \quad (k = 1,2,\ldots N_m) \end{aligned} \qquad \text{(VII.5.22)}$$

Lower and upper principal representations of the given moments μ_k can now be built up; for instance, in the case $N_m = 2q$, the lower principal representation of the moments μ_k is defined by the set of abscissae $u_j^{(L)}$ and weights $W_j^{(L)}$ which satisfy the equations

$$\mu_k = \sum_{j=1}^{q} W_j^{(L)}\, u_j^{(L)} \qquad (k = 1,2,\ldots 2q) \qquad \text{(VII.5.23)}$$

while the upper principal representation of the same moments μ_k is given by the set of abscissae $u_j^{(U)}$ and weights $W_j^{(U)}$ which satisfy the equations

$$\mu_k = \sum_{j=1}^{q+1} W_j^{(U)}\, u_j^{(U)} \qquad \text{(VII.5.23')}$$

$u_1^{(U)}$ being fixed at a, the lower limit of the integral in eq. (VII.5.20) and $u_{q+1}^{(U)}$ fixed at the upper limit b. The importance of the lower and upper principal representations lies in their outstanding feature of providing rigorous lower and upper bounds, respectively, on the unknown moments of the distribution function $F(u)$ (this prerogative is actually restricted to so-called Tchebycheff systems of moment-generation functions $f_k(u)$ [278]. For the present purposes, we are satisfied with asserting that the set of moment-generating functions $\{f_0(u) = 1, f_1(u) = u, f_k(u) = u^2/(1-\omega_{k-1}^2 u) \quad (k = 2,3,\ldots N_m)\}$ forms a Tchebycheff system of order N_m on $[0,E_{10}^{-2}]$ and that the same is true for the extended system $\{f_k(u), u^2/(1+\omega^2 u)\}$).

The practical problem of moment theory involves the solution of the set of equations

$$\frac{4}{3}\pi N \langle\varphi_0|\sum_{i=1}^{N} \delta(\underline{r}_i)|\varphi_0\rangle = \sum_j W_j$$

$$N = \sum_j W_j u_j \qquad \text{(VII.5.24)}$$

$$\alpha(\omega_k) = \sum_j W_j \frac{u_j^2}{1-\omega_k^2 u_j} \qquad (k = 2,\ldots N_m)$$

for weights W_j and abscissae u_j belonging to the two principal representations, the successive construction of the quantity $\sum_j W_j u_j^2/(1+\omega_i^2 u_j)$ will provide rigorous lower and upper bounds to $\alpha(i\omega_k)$ for any ω_k value. If we then solve for oscillator strengths f_{no} and transitions frequencies E_{no} belonging to the two principal representations,

$$f_{no} \to W_j u_j \quad , \quad E_{no} \to u_j^{-\frac{1}{2}} \qquad \text{(VII.5.25)}$$

the dispersion energy coefficient C_6^{disp} can be calculated directly from eq. (VII.4.1).

A closely related moment theory approach [272-75] uses as input data $\frac{4}{3}\pi N\langle\varphi_0|\sum_{j=1}^{N}\delta(\underline{r}_j)|\varphi_0\rangle$, N, the static polarizability $\alpha(0)$ and a number of higher Cauchy moments (see eq. C.5) $\mu_{k+2} = [1/(2k)!][d^{2k}\alpha(\omega)/d\omega^{2k}]_{\omega=0}$, accessible from refractive index or Verdet coefficient data [229]. Bounds to $\alpha(i\omega)$, eq. (VII.5.16) follow by evaluating the involved integral by Gaussian quadrature, i.e.

$$\alpha(i\omega) = \sum_j W_j \frac{u_j^2}{1+\omega_j^2 u_j} + R$$

the quantity R being an (unknown) remainder: abscissae u_j and weights W_j can, in fact, be chosen so as to warrant that the remainder R is either positive or negative definite, thereby providing lower and upper bounds, respectively, to $\alpha(i\omega)$. The moment theory approach along the lines now sketched has been proved to be equivalent to a Padé approximant procedure which makes use of the same amount of information input [184].

The set of nonlinear equations (VII.5.24) can be solved by a number of procedures [278-80]. Out of these, linear programming appears a particularly interesting approach [281], because it allows to increase the flexibility of moment theory method: among other things, we are, for example, enabled to account for the experimental nature of the known moments, introducing in the algorithm their experimental uncertainties; at the same time, the knowledge of discrete and continuum oscillator strength values can be explicitly exploited.

Several dipole-dipole dispersion energy coefficients C_6^{disp} evaluated in terms of Padé approximants or moment theory are collected in Table VII.6. The relevant information needed for their evaluation comes from experimental sources, except for the left-hand side quantity of eq. (VII.5.19). For the interactions involving molecular partners, the values reported for C_6^{disp} correspond to rotationally averaged quantities. The substantial narrowing of the uncertainty range on passing from the [1,0] to the [2,1] Padé approximant is obvious, along with the close agreement between [2,1] Padé approximant and moment theory results. We point out also the agreement between dispersion coefficients of Table VII.6 and those reported in Table VII.5, in the few cases where comparisons are possible.

There exist various other dispersion force bounding procedures [268,269,282-88], which will be left out of the present review. A concise discussion of the relations of some of them with the Padé approximant approach can be found in ref. [184], while for a unified presentation of different formulations, in terms of inner projection techniques, ref. [289] stands as a valid and elegant paper for any reader interested in pushing further the study of the subject of the present section. As far as we are concerned, we shall put an end to this lengthy

Table VII.6. Dipole-dipole dispersion energy coefficients C_6^{disp} (a.u.) for several interacting partners, evaluated by some bounding procedures. The quantities reported correspond to the arithmetic mean of upper and lower bound values.

Interaction	Padé approximants[a] [1,0]	Padé approximants[a] [2,1]	moment theory[b] + linear programming	inner projection[d] (I) (see text)	inner projection[d] (II) (see text)	moment theory[c]
He - He	1.54±0.19	1.46±0.02	1.4618±0.0004	1.51±0.12	1.47±0.03	1.46±0.025
He - Ne	3.37±0.74	3.16±0.12	3.059±0.083			3.05±0.17
He - Ar	11.4±2.8	9.98±0.5	9.86±0.16			9.90±0.61
Ne - Ne	7.72±2.57	7.03±0.56	6.55±0.87	8.08±2.49	7.24±1.21	6.56±0.72
Ne - Ar	25.5±8.6	21.3±0.8	20.47±0.78			20.7±2.1
Ar - Ar	87.9±30	69.3±5.7	67.7±2.0	75.9±14.6	70.2±5.9	68.9±6.3
He - H_2	4.17±0.38	4.02±0.03	4.005±0.045			
Ne - H_2	8.73±1.36	8.27±0.19	8.10±0.			
Ar - H_2	31.5±5.7	28.3±0.9	28.1±0.8			
H_2 - H_2	12.6±1.	12.1±0.1	12.08±0.33			

(a) Ref. [230]; (b) Ref. [278]; (c) Ref. [272]; (d) Ref. [290].

section by taking a quick look through the merits of the inner projection approach reviewed in sec. III.6.

If we apply the formalism introduced in sec. III.6 to establish bounds for the dispersion energy coefficients, some noticeable simplifications can be carried out in the basic formulae, thanks to the possibility of relaxing the complications associated with the presence of symmetry forcing projectors [see the paragraphs successive to eq. (III. 6.11)]. Our starting point is offered by eq. (VII.1.16) for the average 2-nd order dispersion energy. If we limit ourselves to consider the interaction between two spherically symmetric ground state atoms, the rotational average becomes insignificant and, after using the Wigner-Eckart theorem, we are easily led to the following compact expression for the general dispersion energy coefficient [144,290]

$$C^{\text{disp}}_{2(L_a+L_b+1)} = -\frac{(2L_a+2L_b)!}{(2L_a)!\,(2L_b)!} \cdot \langle\varphi_o^A\varphi_o^B|\hat{Q}^o_{L_a}\hat{Q}^o_{L_b}\hat{R}^{AB}_o\hat{Q}^o_{L_a}\hat{Q}^o_{L_b}|\varphi_o^A\varphi_o^B\rangle \qquad \text{(VII.5.26)}$$

where $(L_a,L_b) = 1,2,\ldots$ and $\hat{Q}^o_L$ is the zero-th component of the L-rank electronic multipole moment tensor operator, eq. (VII.1.7) [the subscripts a and b label quantities relative to the partners A and B, respectively]. If we make use of the results expressed by eqs. (III.6. 8) and (III.6.11), the matrix element appearing in eq. (VII.5.26) can be bounded in the following way

$$\begin{aligned} E_{01}^{-1}\{&\langle\varphi_o^A\varphi_o^B|(\hat{Q}^o_{L_a}\hat{Q}^o_{L_b})^2|\varphi_o^A\varphi_o^B\rangle \\ &- \langle\varphi_o^A\varphi_o^B|\hat{Q}^o_{L_a}\hat{Q}^o_{L_b}(\hat{H}_o-E_1)|\mathbb{k}\rangle\langle\mathbb{k}|(\hat{H}_o-E_1)(\hat{H}_o-E_o)|\mathbb{k}\rangle^{-1}\langle\mathbb{k}|(\hat{H}_o-E_1) \\ &\cdot\hat{Q}^o_{L_a}\hat{Q}^o_{L_b}|\varphi_o^A\varphi_o^B\rangle\} \le \langle\varphi_o^A\varphi_o^B|\hat{Q}^o_{L_a}\hat{Q}^o_{L_b}\hat{R}^{AB}_o\hat{Q}^o_{L_a}\hat{Q}^o_{L_b}|\varphi_o^A\varphi_o^B\rangle \\ &\le \langle\varphi_o^A\varphi_o^B|\hat{Q}^o_{L_a}\hat{Q}^o_{L_b}|\mathbb{h}\rangle\langle\mathbb{h}|E_o-\hat{H}_o|\mathbb{h}\rangle^{-1}\langle\mathbb{h}|\hat{Q}^o_{L_a}\hat{Q}^o_{L_b}|\varphi_o^A\varphi_o^B\rangle \end{aligned} \qquad \text{(VII.5.27)}$$

where $\hat{H}_o \equiv \hat{H}^A_o + \hat{H}^B_o$, $E_o(E_1)$ is the ground state (first excited state) energy of the Hamiltonian operator $\hat{H}_o$ and $E_{01} = E_o - E_1$. $|\mathbb{h}\rangle \equiv \{|h_1\rangle,$

$\ldots, |h_n\rangle\}$ and $|\mathbb{k}\rangle \equiv \{|k_1\rangle, \ldots, |k_n\rangle\}$ are arrays of arbitrary basis vectors, subject only to the restriction that each of them lies in the space of the orthogonal complement to $|\varphi_0^A \varphi_0^B\rangle$. A natural choice for $|\mathbb{h}\rangle$ (and $|\mathbb{k}\rangle$) is in the form $|\mathbb{h}\rangle \equiv \{|h_1^A\rangle|h_1^B\rangle, \ldots, |h_i^A\rangle|h_j^B\rangle, \ldots\}$, i.e. as a product of basis vectors relating to the two atoms, such that $\langle\varphi_0|h_i\rangle = 0$, $\forall_i$. If the still arbitrary basis is chosen for each atom as [51,53,54, 290]

$$|h_i^C\rangle = (\hat{R}_0^C)^p \, \hat{Q}_{L_c}^0 \, |\varphi_0^C\rangle \tag{VII.5.28}$$

where p is an integer at our disposal and $\hat{R}_0^C$ the reduced resolvent for the atom C, <u>all</u> the matrix elements involving atoms A and B in eq. (VII. 5.27) can be expressed in the form of ground state <u>sum rule values</u> of the unperturbed atoms. The simplest choice of this kind is $|\mathbb{h}\rangle = |\mathbb{k}\rangle \equiv |h_1^A\rangle|h_1^B\rangle$, with $|h_1^C\rangle = \hat{R}_0^C \, \hat{Q}_{L_c}^0 |\varphi_0\rangle$. Some annoying but straightforward algebra then leads to the following disequalities

$$\frac{1}{4} \frac{(2L_a+2L_b)!}{(2L_a)!\,(2L_b)!} \frac{S_{L_a}(-2)\; S_{L_b}(-2)}{\dfrac{S_{L_a}(-3)}{S_{L_a}(-2)} + \dfrac{S_{L_b}(-3)}{S_{L_b}(-2)}} \le C^{disp}_{2(L_a+L_b+1)} \tag{VII.5.29}$$

$$\le \frac{1}{4} \frac{(2L_a+2L_b)!}{(2L_a)!\,(2L_b)!} E_{10}^{-1} \left\{ S_{L_a}(-1)\; S_{L_b}(-1) - \frac{M^2}{N} \right\}$$

where we have put

$$M = S_{L_a}(-1)S_{L_b}(-2) + S_{L_a}(-2)S_{L_b}(-1) - E_{10} S_{L_a}(-2)S_{L_b}(-2)$$

$$N = S_{L_a}(-1)S_{L_b}(-3) + S_{L_a}(-3)S_{L_b}(-1) + 2S_{L_a}(-2)S_{L_b}(-2) \tag{VII.5.30}$$

$$- E_{10}\left[S_{L_a}(-2)S_{L_b}(-3) + S_{L_a}(-3)S_{L_b}(-2)\right]$$

and

$$S_L(p) = 2\langle\varphi_0|\hat{Q}_L^0[\hat{H}_0 - E_0]^{p+1}\,\hat{Q}_L^0|\varphi_0\rangle \tag{VII.5.31}$$

is a (ground state) sum rule quantity, defined in analogy to eq. (C.11). For L>1, experiments are not a great help for the determination of $S_L(p)$ values, while provide us with information in the case L = 1. Useful approximations to S(p) can be obtained from the general sum-rule expression

$$S(p) = \sum_j f_j \omega_j^p$$

that can be built up, for example, from effective oscillator strengths f_j and transition frequencies ω_j, or the formula

$$S(p) = S(0)\left[a + \frac{b}{(2.5-p)} + \frac{c}{(2.5-p)^2} + \frac{d}{(2.5-p)^3}\right]^p$$

which is claimed being able to provide reliable approximations to sum-rule values [232], once the coefficients a,b,c,d are chosen in such a way to reproduce known Cauchy moments $\alpha_p = S(-2p-2)$, $p = 0,1,2,\ldots$ (see eq. (C.6)).

The values for the dispersion energy coefficients C_6^{disp} under the heading "inner projection" (I) in Table VII.6 are the arithmetic mean of the bounds obtained from eq. (VII.5.29). The quantities collected in the same Table under (II), which are clearly characterized by tighter bounds, have been evaluated along lines quite similar to those leading to eq. (VII.5.29) in terms of the more extended basis $|\mathbb{h}\rangle = |\mathbb{K}\rangle \equiv \{|h_1^A\rangle|h_1^B\rangle, |h_2^A\rangle|h_2^B\rangle\}$, with $|h_1^C\rangle = \hat{R}_0^C \hat{Q}_L^0|\varphi_0\rangle$, $|h_2^C\rangle = (\hat{R}_0^C)^2\hat{Q}_L^0|\varphi_0\rangle$. Inner projection techniques have also been used for bounding values of the dynamic polarizability on the imaginary frequency axis $\alpha(i\omega)$ [51]; it is of interest to stress that the dispersion energy coefficients resulting from their utilization in the Casimir-Polder formula appear of worse quality with respect to those obtained through the more straightforward procedure here considered [290].

VII.6. Nonexpanded Form of the 2-nd Order Polarization Energy.

Although it is a common practice to approximate the 2-nd order polarization energy by its expanded form in terms of the induction and dispersion energy coefficients discussed in the preceding sections, we should be aware of the nature of such approximation and its limits, so as to develop some critical attitude toward a too undiscriminating use of it.

From sec. (V.3), for two interacting partners A and B the nonexpanded form of the 2-nd order polarization energy has the following structure

$$E_{02} = E^{A}_{02,ind} + E^{B}_{02,ind} + E_{02,disp}$$

where

$$E^{A}_{02,ind} = \langle \varphi^{A}_{0} | (\hat{V}^{A}_{en} + \langle \varphi^{B}_{0} | \hat{V}^{AB}_{ee} | \varphi^{B}_{0} \rangle) \; \hat{R}^{A}_{0} (\hat{V}^{A}_{en} + \langle \varphi^{B}_{0} | \hat{V}^{AB}_{ee} | \varphi^{B}_{0} \rangle) | \varphi^{A}_{0} \rangle$$

$$E^{B}_{02,ind} = \langle \varphi^{B}_{0} | (\hat{V}^{B}_{en} + \langle \varphi^{A}_{0} | \hat{V}^{AB}_{ee} | \varphi^{A}_{0} \rangle) \; \hat{R}^{B}_{0} (\hat{V}^{B}_{en} + \langle \varphi^{A}_{0} | \hat{V}^{AB}_{ee} | \varphi^{A}_{0} \rangle) | \varphi^{B}_{0} \rangle$$

$$E_{02,disp} = \langle \varphi^{A}_{0} \varphi^{B}_{0} | \hat{V}^{AB}_{ee} \; \hat{R}^{AB}_{0} \; \hat{V}^{AB}_{ee} | \varphi^{A}_{0} \varphi^{B}_{0} \rangle$$

Obviously, for two <u>identical</u> subsystems in the <u>same</u> state, $E^{A}_{02,ind} = E^{B}_{02,ind}$.

A really exhaustive knowledge of the behaviour of the 2-nd order polarization energy in nonexpanded form as a function of the relative configuration of the interacting partners A,B is unfortunately lacking, information being essentially restricted to a few simple interactions like $H\cdots H^{+}$ (already considered in sec. V.2), $H\cdots H$, $H\cdots He$, $He\cdots He$ [82,83,92,147,150,291-93].

For two ground state hydrogen atoms, the induction energy has been evaluated in an exact way [294]. If we associate the electron 1 with the atom A centered at the origin and the electron 2 with the atom B whose nucleus is specified with respect to A by the vector $\underline{R}$, then

$$\hat{V}^{A}_{en}(1) + \langle 1s_B(2) | \hat{V}^{AB}_{ee}(1,2) | 1s_B(2) \rangle$$

$$= -\frac{1}{|\underline{r}_1-\underline{R}|} + \langle 1s_B(2) | \frac{1}{r_{12}} | 1s_B(2) \rangle = -\left(1+\frac{1}{|\underline{r}_1-\underline{R}|}\right) e^{-|\underline{r}_1-\underline{R}|} \qquad \text{(VII.6.1)}$$

as easily verified.

The evaluation of $E^{A}_{02,ind}$ can be reduced to that of the matrix element $-\langle 1s_A|(1+\frac{1}{|\underline{r}-\underline{R}|})e^{-|\underline{r}-\underline{R}|}|\varphi_A^{ind}\rangle$, where $|\varphi_A^{ind}\rangle$ is solution of the inhomogeneous differential equation

$$(\hat{H}_o^A - E_o^A)|\varphi_A^{ind}\rangle = \Big[(1+\frac{1}{|\underline{r}-\underline{R}|})\, e^{-2|\underline{r}-\underline{R}|} - \langle 1s_A|(1+\frac{1}{|\underline{r}-\underline{R}|})\, e^{-2|\underline{r}-\underline{R}|}|1s_A\rangle\Big]|1s_A\rangle \qquad \text{(VII.6.2)}$$

(see sec. V.1). This equation can be solved by Green's function technique if we first expand the potential $(1+|\underline{r}-\underline{R}|^{-1})\, e^{-2|\underline{r}-\underline{R}|}$, in terms of Legendre polynomials, as $\sum_{\ell=0}^{\infty} u_\ell(r,R)P_\ell(\cos\theta)$ and then assume for $\varphi_A^{ind}(\underline{r})$ the expansion $\varphi_A^{ind}(\underline{r}) = \sum_{\ell=0}^{\infty} \eta_\ell(r,R)P_\ell(\cos\theta)$.

Table VII.7. 2-nd order induction energy for two interacting ground state hydrogen atoms as a function of their distance R.

R(a.u.)	$-E_{02,ind}$(a.u.)	$\langle\hat{H}\rangle_{HL} + E_{02,ind}$(a.u.)	$(\Delta E)_{exact}$(a.u.)
1.	0.1294	-0.1259	-0.1245
1.2	0.9370(-1)	-0.1657	-0.1649
1.4	0.6802(-1)	-0.1735	-0.1745
1.6	0.4933(-1)	-0.1650	-0.1686
1.8	0.3564(-1)	-0.1487	-0.1551
2.	0.2558(-1)	-0.1291	-0.1381
3.	0.4328(-2)	-0.4600(-1)	-0.5731(-1)
4.	0.6307(-3)	-0.1192(-1)	-0.1637(-1)
6.	0.1154(-4)	-0.6246(-4)	-0.8150(-3)

$N(-a) = N\times 10^{-a}$

The values for the exact induction energy reported in the second column of Table VII.7 should be contrasted with $E_{02,ind} = 0$ at any distance as predicted by the long-range approximation, a consequence of the fact that for spherical atoms each permanent multipole moment vanishes. The importance of the induction energy can be appreciated from the last two columns of the same Table, where the exact interaction energy $(\Delta E)_{exact}$ for the $X^1\Sigma_g^+$ state of the hydrogen molecule is compared with $\langle\hat{H}\rangle_{HL} + E_{02,ind}$, the sum of the Heitler-London energy and the 2-nd order induction energy. The noticeable role of the induction energy in the region around the minimum, where we are faced with the strong interaction leading to chemical bonding, is apparent: it is however difficult to assert to what extent this result can quantitatively be extrapolated to more general situations.

An <u>exact</u> knowledge of the behaviour of the dispersion energy is <u>not</u> available even for the case of two hydrogen atoms. A differential equation approach, entirely analogous to that used previously for treating the induction energy, is formally possible also here and reduces the evaluation of the dispersion energy to the matrix element $\langle 1s_A(1)1s_B(2)|(r_{12})^{-1}|\varphi^{disp}(1,2)\rangle$, $|\varphi^{disp}(1,2)\rangle$ being solution to the differential equation

$$[\hat{H}_A^0(1) + \hat{H}_B^0(2) - (E_A^0 + E_B^0)]|\varphi^{disp}(1,2)\rangle$$
$$= -[\hat{1}-\hat{O}_0^A][\hat{1}-\hat{O}_0^B](r_{12})^{-1}|1s_A(1)1s_B(2)\rangle \tag{VII.6.3}$$

where $\hat{O}_0^A = |1s_A(1)\rangle\langle 1s_A(1)|$, $\hat{O}_0^B = |1s_B(2)\rangle\langle 1s_B(2)|$.

Since an exact solution to this equation has not been given, one must be satisfied with approximations founded on the minimization of the following Hylleraas-type functional (see sec. V.1)

$$J[\tilde{\varphi}] = \langle\tilde{\varphi}(1,2)|\hat{H}_0^A(1) + \hat{H}_0^B(2) - (E_A^0 + E_B^0)|\tilde{\varphi}(1,2)\rangle$$
$$+ 2\langle\tilde{\varphi}(1,2)|(r_{12})^{-1}|1s_A(1)1s_B(2)\rangle \tag{VII.6.4}$$

with respect to parameters contained in the trial function $|\tilde{\varphi}(1,2)\rangle$, orthogonal in strong sense to both $|1s_A(1)\rangle$ and $|1s_B(2)\rangle$, so that $\langle 1s_A(1)|\tilde{\varphi}(1,2)\rangle = \langle 1s_B(2)|\tilde{\varphi}(1,2)\rangle = 0$. It is important to recall that $J[\tilde{\varphi}] \geq E_{02,disp}$

If we now use for $(r_{12})^{-1}$ its bipolar expansion as given by eq.

(VI.2.7), and next expand $|\tilde{\varphi}(1,2)\rangle$ in the form

$$|\tilde{\varphi}(1,2)\rangle = \sum_{L_a=0}^{\infty}\sum_{L_b=0}^{\infty}\left(\sum_{M=-L_<}^{+L_<}|\tilde{\varphi}_{L_a,L_b,M}(1,2)\rangle\right)$$

$$\equiv \sum_{L_a=0}^{\infty}\sum_{L_b=0}^{\infty}|\tilde{\varphi}_{L_a,L_b}(1,2)\rangle$$

it is not difficult to show that the approximate dispersion energy $\tilde{E}_{02,disp}$ can be expressed as

$$\tilde{E}_{02,disp} = \sum_{L_a=0}^{\infty}\sum_{L_b=0}^{\infty}\tilde{E}^{disp}_{L_a,L_b}$$

where $\tilde{E}^{disp}_{L_a,L_b} = \langle\tilde{\varphi}_{L_a,L_b}(1,2)|(r_{12})^{-1}|1s_A(1)1s_B(2)\rangle$.

Dispersion energies for two ground state hydrogen atoms as a function of the distance R are reported in Table VII.8. We note that the spherical contribution $E^{disp}_{0,0}$ to $E_{02,disp}$, something which is found only when overlap effects are taken into account, is a rather small quantity over the whole range of interatomic separations. The non-expanded dipole-dipole and dipole-quadrupole contributions $E^{disp}_{1,1}$ and $E^{disp}_{1,2}+E^{disp}_{2,1} = 2E^{disp}_{1,2}$ to $E_{02,disp}$ are compared with the corresponding long-range approximations $-C_6R^{-6}$ and C_8R^{-8}. The completely unreliable estimate for the dispersion energy as given by the long-range approximation at short distances is immediately recognized, along with the progressive adequacy of such approximation as the distance R becomes sufficiently large.

As a very brief comment on the nature of the calculations which the entries reported in Table VII.8 are based upon, it is not useless to emphasize that in most cases the trial functions $\tilde{\varphi}_{L_a,L_b,M}(1,2)$ were chosen in the form of a single pseudo-state, to be optimized at each distance with respect to the screening constant of the involved orbitals. Such a simple procedure, which has already been shown to lead to rather accurate values for multipole polarizabilities of simple systems, appears really gratifying in view of its possible extension to more complex systems.

Before ending this section, we should like to describe an elegant

Table VII.8. Dispersion energy (a.u.) for two ground state hydrogen atoms as a function of the distance R.

R(a.u.)		$-\tilde{E}^{disp}_{0,0}$	$-\tilde{E}^{disp}_{1,1}$	$-C_6R^{-6}$	$-2\tilde{E}^{disp}_{1,2}$	$-C_8R^{-8}$
	Ref. [92]		0.4$\underline{365}$(-2)	0.1015(0)	0.2$\underline{074}$(-2)	0.4859(0)
2.	Ref. [150]		0.3943(-2)		0.1962(-2)	
	Ref. [293]	0.8713(-4)	0.3518(-2)			
	Ref. [291]		0.5712(-3)		0.221 (-3)	
4.	Ref. [92]		0.72$\underline{26}$(-3)	0.1587(-2)	0.34$\underline{48}$(-3)	0.1898(-2)
	Ref. [150]		0.6747(-3)		0.3371(-3)	
	Ref. [293]	0.5551(-5)	0.6951(-3)			
	Ref. [291]		0.1146(-3)		0.418 (-4)	
6.	Ref. [92]		0.120$\underline{8}$(-3)	0.1393(-3)	0.48$\underline{99}$(-4)	0.7406(-4)
	Ref. [150]		0.1170(-3)		0.4605(-4)	
	Ref. [293]	0.2240(-6)	0.1175(-3)			
	Ref. [291]		0.24 (-4)		0.66 (-5)	
8.	Ref. [92]		0.2436(-4)	0.2479(-4)	0.6950(-5)	0.7415(-5)
	Ref. [150]		0.2404(-4)		0.6692(-5)	
10.	Ref. [92]		0.6489(-5)	0.6499(-5)	0.1236(-5)	0.1244(-5)
	Ref. [150]		0.6432(-5)		0.1218(-5)	

The underlined figures are to be considered uncertain (unstable with respect to an extension of the basis set employed in the calculation). The notation N(-a) means $N\times 10^{-a}$.

approach to the evaluation of dispersion (and induction) energies without neglecting overlap effects, which has until now received a moderate consideration [295-98]. The dispersion energy can be expressed in the form [131,176,299]

$$E_{02,disp} = \iiiint d\underline{r}_1 d\underline{r}_2 d\underline{r}_1' d\underline{r}_2' \quad \text{(VII.6.5)}$$

$$\frac{\langle \varphi_0^A \varphi_0^B | \hat{\rho}_A^{e\ell}(\underline{r}_1) \hat{\rho}_B^{e\ell}(\underline{r}_2) \hat{R}_0^{AB} \hat{\rho}_A^{e\ell}(\underline{r}_1') \rho_B^{e\ell}(\underline{r}_2') | \varphi_0^A \varphi_0^B \rangle}{|\underline{R}+\underline{r}_2-\underline{r}_1|\;|\underline{R}+\underline{r}_2'-\underline{r}_1'|}$$

if we use for $\hat{V}_{ee}^{AB}$ its integral representation in terms of the electronic charge density operators, eqs. (VI.1.1), (VI.1.2). It is now convenient to exploit the Fourier representation of the Coulomb potential [300]

$$\frac{1}{|\underline{R}+\underline{r}_2-\underline{r}_1|} = \frac{1}{2\pi^2}\int d\underline{k}\; k^{-2}\exp\left[i\underline{k}\cdot(\underline{R}+\underline{r}_2-\underline{r}_1)\right]$$

for transforming eq. (VII.6.5); we obtain in a direct way

$$E_{02,disp} = \frac{1}{4\pi^4}\iint d\underline{k}\, d\underline{k}'\, k^{-2}k'^{-2}\exp[i(\underline{k}+\underline{k}')\cdot\underline{R}]$$

$$\cdot\iiiint d\underline{r}_1\, d\underline{r}_2\, d\underline{r}_1'\, d\underline{r}_2' \exp[i\underline{k}\cdot(\underline{r}_2-\underline{r}_1)]\exp[i\underline{k}'\cdot(\underline{r}_2'-\underline{r}_1')]$$

$$\cdot\langle\varphi_o^A\varphi_o^B|\hat{\rho}_A^{e\ell}(\underline{r}_1)\hat{\rho}_B^{e\ell}(\underline{r}_2)\hat{R}_o^{AB}\hat{\rho}_A^{e\ell}(\underline{r}_1')\hat{\rho}_B^{e\ell}(\underline{r}_2')|\varphi_o^A\varphi_o^B\rangle \qquad \text{(VII.6.6)}.$$

Let us now introduce Fourier transforms of the charge operators $\hat{\rho}(\underline{r})$:

$$\int d\underline{r}\,\exp(i\underline{k}\cdot\underline{r})\,\hat{\rho}^{e\ell}(\underline{r}) = -\sum_j^{N_e}\exp(i\underline{k}\cdot\underline{r}_j) \equiv \hat{\rho}(\underline{k}) \qquad \text{(VII.6.7)}$$

where N_e is the total number of electrons. Eq. (VII.6.6) can be transformed very easily as follows

$$E_{02,disp} = \frac{1}{4\pi^4}\iint d\underline{k}\, d\underline{k}'\, k^{-2}k'^{-2}\exp[i(\underline{k}-\underline{k}')\cdot\underline{R}]$$

$$\cdot\langle\varphi_o^A\varphi_o^B|\hat{\rho}_A(-\underline{k})\hat{\rho}_B(\underline{k})\hat{R}_o^{AB}\hat{\rho}_A(\underline{k}')\hat{\rho}_B(-\underline{k}')|\varphi_o^A\varphi_o^B\rangle$$

Now the resolvent $\hat{R}_o^{AB} = -\sum_n{}'\sum_m{}'(E_{no}^A+E_{mo}^B)^{-1}|\varphi_n^A\varphi_m^B\rangle\langle\varphi_n^A\varphi_m^B|$, through the simple use of the integral identity (VII.1.18), can be cast into the form

$$\hat{R}_o^{AB} = -\frac{2}{\pi}\int_0^\infty d\omega\left[\sum_n{}' \frac{|\varphi_n^A\rangle\langle\varphi_n^A|E_{no}^A}{(E_{no}^A)^2+\omega^2}\right]\left[\sum_m{}' \frac{|\varphi_m^B\rangle\langle\varphi_m^B|E_{mo}^B}{(E_{mo}^B)^2+\omega^2}\right] \qquad \text{(VII.6.8)}$$

which allows $E_{02,disp}$ to be written compactly as

$$E_{02,disp} = -\frac{1}{8\pi^5}\iiint d\omega\, d\underline{k}\, d\underline{k}' k^{-2} k'^{-2} \exp[i(\underline{k}-\underline{k}')\cdot\underline{R}]$$

$$\cdot\alpha_A(\underline{k},\underline{k}';i\omega)\alpha_B(-\underline{k},-\underline{k}';i\omega) \qquad \text{(VII.6.9)}$$

having introduced the generalized polarizability

$$\alpha(\underline{k},\underline{k}';\omega) = 2\sum_n{}' \frac{E_{no}<\varphi_0|\hat{\rho}(-\underline{k})|\varphi_n><\varphi_n|\hat{\rho}(\underline{k}')|\varphi_0>}{E_{no}^2-\omega^2} \qquad \text{(VII.6.10)}$$

Eq. (VII.6.9) is our final result. If values of the generalized polarizability prolonged in the complex plane to the imaginary axis can be obtained as a function of $\underline{k},\underline{k}'$, one has at disposal an elegant approach for evaluating $E_{02,disp}$. We note incidentally that the matrix element $<\varphi_n|\hat{\rho}(\underline{k})|\varphi_0>$ is the same encountered in the theory of the inelastic scattering of charged particles by atoms or molecules in the Born Approximation [301]; some interesting results arising from an analysis of such matrix element in the complex plane [302-4] have recently been used for approximating dispersion energies of closed-shell atoms [296].

The most serious problem posed by the approach under investigation appears the evaluation of $\alpha(\underline{k},\underline{k}';\omega)$. A simple manipulation of eq. (VII. 6.10) founded on the identity $2E_{no}[(E_{no})^2-\omega^2]^{-1} = (E_{no}+\omega)^{-1} + (E_{no}-\omega)^{-1}$ allows to write

$$\alpha(\underline{k},\underline{k}';\omega) = \alpha_+(\underline{k},\underline{k}';\omega) + \alpha_-(\underline{k},\underline{k}';\omega)$$

where

$$\alpha_\pm(\underline{k},\underline{k}';\omega) = \sum_n{}' \frac{<\varphi_0|\hat{\rho}(-\underline{k})|\varphi_n><\varphi_n|\hat{\rho}(\underline{k}')|\varphi_0>}{E_{no}\pm\omega} \qquad \text{(VII.6.11)}$$

One should also note that $\alpha_-(\underline{k},\underline{k}';\omega) = \alpha_+(\underline{k},\underline{k}';-\omega)$. The infinite summation involved in the definition of $\alpha_+(\underline{k},\underline{k}';\omega)$, eq. (VII.6.11), is in principle avoided if one is able to solve the following differential equation

$$(\hat{H}_0-E_0+\omega)|\phi(\underline{k}',\omega)> = [\hat{\rho}(\underline{k}')-<\varphi_0|\hat{\rho}(\underline{k}')|\varphi_0>]|\varphi_0> \qquad \text{(VII.6.12)}$$

As a matter of fact, it is easily verified that

$$\alpha_+(\underline{k},\underline{k}';\omega) = \langle\varphi_0|\hat{\rho}(-\underline{k})|\phi(\underline{k}',\omega)\rangle$$

However, it is much more likely that we are only able to approximately solve eq. (VII.6.12)* ; for instance, sensible approximations $|\tilde{\phi}(\underline{k}',\omega)\rangle$ to the true solution can be obtained from the variation of the functional

$$J[\tilde{\phi}] = \langle\tilde{\phi}(\underline{k}';\omega)|\hat{H}_0 - E_o + \omega|\tilde{\phi}(\underline{k}';\omega)\rangle$$
$$- \langle\tilde{\phi}(\underline{k}';\omega)|\hat{\rho}(\underline{k}')|\varphi_o\rangle - \langle\varphi_o|\hat{\rho}^\dagger(\underline{k}')|\tilde{\phi}(\underline{k}';\omega)\rangle \qquad \text{(VII.6.13)}$$

with the restriction that $\langle\varphi_0|\tilde{\phi}(\underline{k}';\omega)\rangle = 0$ [295]. Applications of this procedure are actually in demand.

VII.7. Approximate Evaluation of Exchange Polarization Effects.

The importance of taking into account polarization effects in molecular interaction energy calculations should have clearly been appreciated in the course of these notes. Exchange polarization contributions complement the Coulomb polarization energy in a measure not easily assessable, even though, from the simple inspection of Table V.2 for the case of the interaction $H\cdots H^+$, one is led to surmise a not negligible role of these exchange effects. The 2-nd order exchange polarization energy [see eqs. (V.3.20), (V.3.21)] appears a really hard quantity to be evaluated, more exacting than the Coulomb polarization energy, in view of the fact that indirect or semiempirical estimates do not seem to have got a footing; being frequently ignored in actual calculations, we think it right to reserve a few pages to a (concise) review of its role.

Before going into details, we recall that 2-nd and higher order contributions to the exchange energy depend upon the formalism chosen

---

*Some interesting results concerning the generalized polarizability of the ground state hydrogen atom can be found in Refs. [305,306].

for forcing from the outset the required symmetry in the perturbation theory (see Ch.III). We have also remarked how the MS-MA procedure (see sec. V.1) appears a valuable and viable approach, one of its merit lying in the relative simplicity, which allows the evaluation of the total, i.e. Coulomb plus exchange, 2-nd order energy to be carried out only in terms of the solution $|\varphi_{01}\rangle$ to the exchangeless equation $(\hat{H}_0-E_0)|\varphi_{01}\rangle = (1-|\varphi_0\rangle\langle\varphi_0|)\hat{V}|\varphi_0\rangle$ [see eq. (III.4.16')]. We shall confine our attention to such symmetry-adapted procedure and limitedly to closed-shell interacting "monomers", in the approximation of neglecting intracorrelation effects (i.e., only the leading term in a perturbative series of Møller-Plesset type, along the lines of sec. V.4).

The starting point of the present analysis is provide by eqs. (V.3.20), (V.3.21). Once the Møller-Plesset procedure is applied, at the lowest order we find the following approximations to exchange induction and exchange dispersion energy, respectively

$$W_{02,\text{ex-ind}} = \langle\phi^A_{01}\phi^B_0|(\hat{\mathcal{P}}^{AB}-\mathcal{P}^{AB}_0)(\hat{V}-W_{01})|\phi^A_0\phi^B_0\rangle + \langle\phi^A_0\phi^B_{01}|(\hat{\mathcal{P}}^{AB}-\mathcal{P}^{AB}_0)(\hat{V}-W_{01})|\phi^A_0\phi^B_0\rangle \tag{VII.7.1}$$

$$W_{02,\text{ex-disp}} = \langle\phi^{AB}_{01}|(\hat{\mathcal{P}}^{AB}-\mathcal{P}^{AB}_0)(\hat{V}-W_{01})|\phi^A_0\phi^B_0\rangle \tag{VII.7.2}$$

$|\phi^A_{01}\rangle$ and $|\phi^B_{01}\rangle$ are solutions to the "one-center" equations

$$[W^A_0-\hat{F}^A_0]|\phi^A_{01}\rangle = \hat{P}^A_0[\hat{V}^A_{en}+\langle\phi^B_0|\hat{V}^{AB}_{ee}|\phi^B_0\rangle]|\phi^A_0\rangle$$
$$[W^B_0-\hat{F}^B_0]|\phi^B_{01}\rangle = \hat{P}^B_0[\hat{V}^B_{en}+\langle\phi^A_0|\hat{V}^{AB}_{ee}|\phi^A_0\rangle]|\phi^B_0\rangle \tag{VII.7.3}$$

while $|\phi^{AB}_{01}\rangle$ is determined by the "two-center" equation

$$[(W^A_0-\hat{F}^A_0)+(W^B_0-\hat{F}^B_0)]|\phi^{AB}_{01}\rangle = \hat{P}^A_0\hat{P}^B_0\hat{V}^{AB}_{ee}|\phi^A_0\rangle|\phi^B_0\rangle \tag{VII.7.4}$$

$|\phi^A_0\rangle$ and $|\phi^B_0\rangle$ denote Slater determinants corresponding to the H.F. ground states of the partners A and B, respectively, solutions to the equations $(W^A_0-\hat{F}^A_0)|\phi^A_0\rangle = 0$, $(W^B_0-\hat{F}^B_0)|\phi^B_0\rangle = 0$; $\hat{V} = \hat{V}^A_{en}+\hat{V}^B_{en}+\hat{V}^{AB}_{ee}$ is the interaction potential energy operator [see eq. (V.3.12)] and $\hat{P}^C_0$ is the projector onto the orthogonal complement to $|\phi_0^C\rangle$, while W_{01} is the (approximate) electrostatic contribution to the 1-st order interaction energy, $W_{01} = \langle\phi_0^A\phi_0^B|\hat{V}|\phi_0^A\phi_0^B\rangle$.

Eqs. (VII.7.3), (VII.7.4) can be simplified if we consider that $|\phi_0^C\rangle = (N^C!)^{\frac{1}{2}}\hat{A}^C|w_0^C\rangle$, [see eq. (V.4.2)], where N^C is the number of electrons belonging to the "monomer" C, $\hat{A}^C \equiv (N^C!)^{-1}\sum_{j_C}(-1)^{p_{j_C}}\hat{P}_{j_C}$ is the antisymmetrizing operator (a projector) associated with C [see eq. (II.3.2)] and $|w_0^C\rangle$ a Hartree product of the spinorbitals occupied in the ground state of C. Since $[\hat{A}^C,\hat{F}_o^C] = [\hat{A}^C,\hat{V}_{en}^C + \langle\phi^D|\hat{V}_{ee}^{CD}|\phi^D\rangle] = 0$, it is straightforward to verify that the solutions to eqs. (VII.7.3), (VII.7.4) can be expressed in the form

$$|\phi_{01}^C\rangle = (N^C!)^{\frac{1}{2}}\,\hat{A}^C|u_{01}^C\rangle \tag{VII.7.5}$$

$$|\phi_{01}^{AB}\rangle = (N^A!N^B!)^{\frac{1}{2}}\,\hat{A}^A\hat{A}^B|u_{01}^{AB}\rangle \tag{VII.7.6}$$

where

$$\begin{aligned}
[W_0^A-\hat{F}_0^A]|u_{01}^A\rangle &= [1-(N^A!)^{\frac{1}{2}}|w_0^A\rangle\langle\phi_0^A|][\hat{V}_{en}^A+\langle\phi_0^B|\hat{V}_{ee}^{AB}|\phi_0^B\rangle]|w_0^A\rangle \\
[W_0^B-\hat{F}_0^B]|u_{01}^B\rangle &= [1-(N^B!)^{\frac{1}{2}}|w_0^B\rangle\langle\phi_0^B|][\hat{V}_{en}^B+\langle\phi_0^A|\hat{V}_{ee}^{AB}|\phi_o^B\rangle]|w_o^B\rangle
\end{aligned} \tag{VII.7.7}$$

and

$$\begin{aligned}
[(W_o^A-\hat{F}_o^A)+(W_o^B-\hat{F}_0^B)]|u_{01}^{AB}\rangle &= [1-(N^A!)^{\frac{1}{2}}|w_0^A\rangle\langle\phi_0^A|] \\
&\cdot[1-(N^B!)^{\frac{1}{2}}|w_0^B\rangle\langle\phi_0^B|]\hat{V}_{ee}^{AB}|w_0^A\rangle|w_0^B\rangle
\end{aligned} \tag{VII.7.8}$$

As a simple application of the preceding general equations, we examine here the case of the interaction He···He [105,307], for which $|w_0^A\rangle \equiv |a(1)\rangle|a(2)\rangle|\frac{1}{2}(1)\rangle|-\frac{1}{2}(2)\rangle$, $|w_0^B\rangle \equiv |b(3)\rangle|b(4)\rangle|\frac{1}{2}(3)\rangle|-\frac{1}{2}(4)\rangle$, $|\phi_0^A\rangle = (2)^{-\frac{1}{2}}(1-\hat{P}_{12})|w_0^A\rangle$, $|\phi_0^B\rangle = (2)^{-\frac{1}{2}}(1-\hat{P}_{34})|w_0^B\rangle$, $\hat{V}_{en}^A = -(2/r_{1B}+2/r_{2B})$, $\hat{V}_{ee}^{AB} = 1/r_{13}+1/r_{14}+1/r_{23}+1/r_{24}$ [$|a\rangle|\pm\frac{1}{2}\rangle$ denote the spinorbitals occupied in the ground state of the He atom labelled A; $|b\rangle|\pm\frac{1}{2}\rangle$ have an analogous meaning for the other atom; see eq. (V.4.15)].

The effect of the polarization of inductive type, due to the presence of the neighbouring atom, is contained in $|u_{01}^A(1,2)\rangle$ and $|u_{01}^B(3,4)\rangle$, which can be expressed in the form

$$|u_{01}^{C}(i,j)\rangle = [|\omega^{C}(i)\rangle|c(j)\rangle + |c(i)\rangle|\omega^{C}(j)\rangle] \cdot |\tfrac{1}{2}(i)\rangle| -\tfrac{1}{2}(j)\rangle \quad \text{(VII.7.9)}$$

It is rather straightforward to show from eqs. (VII.7.7), (VII.7.9) that the polarized orbital $|\omega^{A}\rangle$ satisfies the following equation

$$[\varepsilon_{1}^{A} - \hat{h}_{A}^{HF}(1)]|\omega^{A}(1)\rangle = [1-|a(1)\rangle\langle a(1)|] \cdot [-2/r_{1B} + 2\langle b(3)|1/r_{13}|b(3)\rangle]|a(1)\rangle \quad \text{(VII.7.10)}$$

with an entirely analogous equation for $|\omega^{B}(3)\rangle$. The role of the mean electric field generated by the electrons "belonging" to the atom B in determining the inductive correction to $|a(1)\rangle$ is quite manifest in eq. (VII.7.10), according to the general remarks of sec. V.3. In eq. (VII.7.10) ε_{1}^{A} is the one-electron energy associated with the occupied orbital $|a\rangle$, eigenvector to the H.F. equation $\hat{h}_{A}^{HF}|a\rangle = \varepsilon_{1}^{A}|a\rangle$.

$|u_{01}^{AB}\rangle$, solution to eq. (VII.7.8), takes into account dispersion effects; its expression,

$$|u_{01}^{AB}(1,2,3,4)\rangle = [|u^{AB}(1,3)\rangle|a(2)\rangle|b(4)\rangle + |u^{AB}(1,4)\rangle|a(2)\rangle|b(3)\rangle + |u^{AB}(2,3)\rangle|a(1)\rangle|b(4)\rangle + |u^{AB}(2,4)\rangle|a(1)\rangle|b(3)\rangle] \cdot |\tfrac{1}{2}(1)\rangle|-\tfrac{1}{2}(2)\rangle|\tfrac{1}{2}(3)\rangle|-\tfrac{1}{2}(4)\rangle \quad \text{(VII.7.11)}$$

once substituted into eq. (VII.7.8) leads to the following equation for the (spin-free) dispersion pair function $|u^{AB}(i,j)\rangle$ ($i\equiv 1,2$; $j\equiv 3,4$)

$$\{[\varepsilon_{1}^{A} - \hat{h}_{A}^{HF}(i)] + [\varepsilon_{1}^{B} - \hat{h}_{B}^{HF}(j)]\}|u^{AB}(i,j)\rangle = [1-|a(i)\rangle\langle a(i)|] \cdot [1-|b(j)\rangle\langle b(j)|]\frac{1}{r_{ij}}|a(i)\rangle|b(j)\rangle \quad \text{(VII.7.12)}$$

with the strong-orthogonality conditions $\langle a(1)|u^{AB}(1,2)\rangle = \langle b(2)|u^{AB}(1,2)\rangle = 0$. The nature of the perturbing operator $1/r_{ij}$ involved in the equation for the dispersion pair function $|u^{AB}(i,j)\rangle$ should be noted and contrasted with that of the physical perturbation active in determining the <u>induction</u> effect: all this, of course, is perfectly in line with our expectation after the remarks of sec. V.3.

After some annoying algebra, one is led to the following expressions for exchange induction and exchange dispersion energy of two interacting helium atoms, at the lowest order of a Møller-Plesset perturbative procedure [307]

$$W_{02,\text{ex-ind}} = -4\langle ab|\hat{v}|b\omega^A\rangle - 4\langle ab|\hat{v}|bb\rangle\langle b|\omega^A\rangle + 12S^2\langle ab|\hat{v}|\omega^A b\rangle$$
$$+12S\langle ab|\hat{v}|ab\rangle\langle b|\omega^A\rangle - 4S\langle ab|\hat{v}|a\omega^A\rangle - 4S\langle ab|\hat{v}|\omega^A a\rangle$$
$$+O(S^4) \qquad \text{(VII.7.13)}$$

$$W_{02,\text{ex-disp}} = K_1 + K_2 + K_3 + O(S^4) \qquad \text{(VII.7.14)}$$

where

$$K_1 = -2\iint d\underline{r}_1 d\underline{r}_2\, a(1)b(2)v(1,2)u^{AB}(2,1)$$

$$K_2 = -2\iiint d\underline{r}_1 d\underline{r}_2 d\underline{r}_3\, a(1)a(2)b(3)v(1,3)[u^{AB}(1,2)a(3)$$
$$+a(1)u^{AB}(3,2)] - 2\iiint d\underline{r}_1 d\underline{r}_2 d\underline{r}_3\, a(1)b(2)v(1,3)b(3)$$
$$\cdot[u^{AB}(2,1)b(3) + b(1)u^{AB}(2,3)] \qquad \text{(VII.7.15)}$$

$$K_3 = -2S\iiint d\underline{r}_1 d\underline{r}_2 d\underline{r}_3\, a(1)b(3)v(1,3)[u^{AB}(1,2)a(2)b(3)$$
$$+a(1)b(2)u^{AB}(2,3)] + 6S^2\iint d\underline{r}_1 d\underline{r}_2\, a(1)b(2)v(1,2)u^{AB}(1,2)$$
$$+6\langle ab|\hat{v}|ab\rangle\iint d\underline{r}_1 d\underline{r}_2\, a(1)b(2)u^{AB}(2,1)$$

In the preceding equations we have introduced the overlap integral $S = \langle a|b\rangle$ and put

$$\hat{v} \equiv \hat{v}(i,j) \equiv \frac{1}{r_{ij}} - \frac{1}{r_{iB}} - \frac{1}{r_{jA}}$$

while $O(S^4)$ means that (negligible) double-exchange terms have been omitted (see sec. V.3).

The complexity of the results expressed by eqs. (VII.7.13)-(VII.7.15) should be contrasted with the relative simplicity of the Coulomb induction and dispersion energy expressions (in the approximation of neglecting intracorrelation effects), which are obviously obtainable in terms of the quantities $|\omega^A\rangle$ and $|u^{AB}(i,j)\rangle$:

$$W_{02,ind} = 4\langle a|-\frac{2}{r_{1B}} + 2\langle b(3)|\frac{1}{r_{13}}|b(3)\rangle|\omega^A\rangle \qquad (VII.7.16)$$

$$W_{02,disp} = 4\ \langle ab|\hat{v}|u^{AB}\rangle \qquad (VII.7.17)$$

It should be obvious that on expanding in terms of 1/R the quantities $-2/r_{1B} + 2\langle b(3)|1/r_{13}|b(3)\rangle$, $\hat{v}(1,2)$, $|\omega^A\rangle$ and $|u^{AB}\rangle$ in the preceding equations, one is led to approximations for the Coulomb induction and dispersion energy characterized by the neglect of charge overlap as well as intra-correlation effects, just the same result obtained by using uncoupled H.F. perturbation theory (see sec. VII.4).

The practical solution of eqs. (VII.7.10) and (VII.7.12), respectively for the quantities $|\omega^A(1)\rangle$, $|\omega^B(3)\rangle$ and $|u^{AB}(1,3)\rangle$, is most likely carried out by resorting to variation-perturbation procedures, through the minimization of suitable functional forms. For the case here investigated, the functionals minimized by $|\omega^A(1)\rangle$ and $|u^{AB}(1,3)\rangle$ are [307,308]

$$J_{ind}[\tilde{\omega}^A] = \langle\tilde{\omega}^A(1)|\hat{h}_A^{HF}(1)-\varepsilon_1^A|\tilde{\omega}^A(1)\rangle+2\langle\tilde{\omega}^A(1)|\hat{P}_0^A[-\frac{2}{r_{1B}} + 2\langle b(3)|\frac{1}{r_{13}}|b(3)\rangle]|a(1)\rangle \qquad (VII.7.18)$$

$$J_{disp}[\tilde{u}^{AB}] = \langle\tilde{u}^{AB}(1,3)|\hat{h}_A^{HF}(1)+\hat{h}_B^{HF}(3)-\varepsilon_1^A-\varepsilon_1^B|\tilde{u}^{AB}(1,3)\rangle +2\langle\tilde{u}^{AB}(1,3)|\hat{P}_0^A\hat{P}_0^B\hat{v}(1,3)|a(1)b(3)\rangle \qquad (VII.7.18')$$

where $|\tilde{\omega}^A\rangle$ and $|\tilde{u}^{AB}\rangle$ are trial vectors and $\hat{P}_0^A=1-|a\rangle\langle a|$, $\hat{P}_o^B=1-|b\rangle\langle b|$. These trial vectors are most frequently expressed in the form of expansions in terms of H.F. eigensolutions to $\hat{h}_A^{HF}$, $\hat{h}_B^{HF}$, i.e.

$$|\tilde{\omega}^A(1)\rangle = \sum_r |\psi_r^A(1)\rangle C_r^A \quad , \quad |\tilde{u}^{AB}(1,3)\rangle = \sum_{rs}|\psi_r^A(1)\rangle|\psi_s^B(3)\rangle C_{rs}^{AB}$$

with the sums running only over the unoccupied part of the spectrum, so as to assure the proper orthogonality conditions. The results for $W_{02,ind}$, $W_{02,disp}$, $W_{02,ex-ind}$ and $W_{02,ex-disp}$ are expected to be only moderately reliable approximations to the true quantities, because of the complete lack of correlation effects (we are here faced with a situation quite similar to that encountered in connection with the electric polarizability evaluation, see sec. VII.3). A (partial) remedy for this flaw would involve including 1-st order corrections in the perturbation operator $(\hat{H}_0-\hat{F}_0)$ so as to make allowance for lowest order intra-correlation effects; as an alternative approach in the same direction, one can also re-formulate the preceding equations in terms of a zero-th order, separable, Hamiltonian operator other than $\hat{F}_0=\sum_i \hat{h}^{HF}(i)$; the choices $\hat{F}_0 = \sum_i \hat{h}(i)$, $\hat{h}(i)$ being the Kelly-Silverstone-Huzinaga-type Hamiltonian operator [see eq. (V.4.11)], or $\hat{F}_0 = \sum_i \hat{h}^H(i)$, a sum of exchangeless Hartree-type single particle Hamiltonian operators, provide a suitable way for taking into (partial) account correlation effects thanks to the replacement of a $\hat{V}^{N-1}$ electronic potential in lieu of $\hat{V}^N$ appearing in $\hat{h}^{HF}$ [107].

In order to acquire some appreciation of the 2-nd order exchange contributions to the interaction energy, the individual components of the interaction energy between two ground state He atoms at R = 5.6 a.u. (i.e. the distance corresponding to their van der Waals minimum [309]) are reported in Table VII.9

Table VII.9. Contributions to the interaction energy (a.u.) between two ground state He atoms, evaluated at R = 5.6 a.u. (van der Waals minimum) [307].

	$\hat{F}_0 = \sum_i \hat{h}^{HF}(i)$	$\hat{F}_0 = \sum_i \hat{h}^{H}(i)$
$W_{01} + W_{01,ex}$	30.81(-6)	30.81(-6)
$W_{02,ind}$	-0.73(-6)	-0.79(-6)
$W_{02,disp}$	-51.02(-6)	-73.69(-6)
$W_{02,ex-ind}$	0.57(-6)	0.63(-6)
$W_{02,ex-disp}$	1.33(-6)	1.84(-6)

The two columns of values correspond to different choices of the zero-th order Hamiltonian operator $\hat{F}_0$. The very strong dependence of the dispersion energy on such choice is quite evident and reflects the sensitivity of this quantity on the correlation effects [according to our previous discussion, $W_{02,disp}$ and $W_{02,ex-disp}$ values evaluated according to the choice $\hat{F}_0 = \sum_i \hat{h}^H(i)$ should be regarded much more reliable estimates of the correct quantities than the values obtained by the choice $\hat{F}_0 = \sum_i \hat{h}^{HF}(i)$]. We take here the opportunity for stressing the decisive role of the dispersion energy in making the ground state of the system He_2 slightly bound.

The extension of the formalism reviewed in this section to more complex interactions than He•••He has received a very limited amount of attention. Without going into details, we mention to this regard, an important investigation on the interaction between two ground state Be atoms [310], for which explicit formulae of the exchange polarization energy have been derived in the approximation of neglecting intra-correlation effects. Concerning the importance of exchange dispersion energy effects, we limit ourselves to assert that for Be_2 they appear greater than for H_2 or He_2 (a quenching of about 5÷10% of the dispersion energy term in Be_2 compared to about 2% in He_2). It is likely therefore that reliable estimates of intermolecular interaction energies in the region around the van der Waals minimum cannot ignore 2-nd order exchange effects.

VIII. EPILOGUE.

In the last two chapters the reader has been brought into contact with various methods by which the calculation of the 1-st and 2-nd order contributions to the interaction energy between two atoms or molecules has most frequently been carried out.

In the course of the presentation of a fairly abundant material, we had recourse to a number of examples concerning simple interactions, with the main intention of stressing given computational approaches and promoting possible comparisons among different methods. The net result is up to this point a rather disconnected and heterogeneous ensemble of _partial_ data for a number of interactions, while it would be of primary interest being able to ascertain, on the basis of _global_ results, how well perturbation procedures (truncated after the 2-nd order) work when applied to typical problems.

The accomplishment of this object appears the only reasonable way at our disposal for definitely bringing about confidence in the theoretical framework developed in the course of these notes: in such conviction, this last chapter is entirely devoted to a quick review of results concerning a few representative interactions involving atoms and molecules.

a) We shall start by considering the interaction between two ground state _hydrogen atoms_, surely the most classic problem encountered in molecular quantum mechanics and one invariably described from the variational point of view in any text-book of Quantum Chemistry [7,84,85]. A noticeable amount of attention has been addressed to this problem also from the perturbative standpoint [311-318]. The main reason for our interest is that, contrary to the case of the interaction $H\cdots H^+$, the present problem mimics the behaviour of more complex interactions, as a result of the presence of the interelectronic repulsion term in the Hamiltonian perturbation operator. At the same time, complicating features associated with our usually imperfect, knowledge of the exact eigenstates to the Hamiltonian operator of the unperturbed partners (which gives rise to the hard problem of introducing intra-correlation corrections) are lacking, so that a good opportunity is offered for appreciating the validity of the various symmetry-adapted perturbation approaches, without confusing approximations of other kind.

Although it should by now be accepted that the truly fundamental symmetry to be forced in the perturbation scheme is that related to the permutational invariance of the Hamiltonian operator H, further evidence can be brought to such point thanks to the simplicity of the interaction under study, which allows a rather straightforward analysis of the question. If we allot the electrons 1 and 2 to the hydrogen atoms centred at a and b, respectively, the unperturbed Hamiltonian operator corresponding to two non-interacting hydrogen atoms is

$$\hat{H}_0 = (-\frac{1}{2}\nabla_1^2 - \frac{1}{r_{1a}}) + (-\frac{1}{2}\nabla_2^2 - \frac{1}{r_{2b}})$$

with unperturbed ground state wavefunction $\langle 1,2|\varphi_0\rangle \equiv \varphi_0(1,2) \equiv 1s_a(1)\cdot 1s_b(2)$, while the perturbation operator $\hat{V}$ is clearly

$$\hat{V} = -\frac{1}{r_{1b}} - \frac{1}{r_{2a}} + \frac{1}{r_{12}} + \frac{1}{R}$$

The full group G of operators commuting with the total Hamiltonian $\hat{H} = \hat{H}_0 + \hat{V}$ can be represented in the following form of direct group product

$$G = \mathscr{C}_i \otimes \mathscr{C}_{\infty v} \otimes S_2$$

where $\mathscr{C}_i \equiv (\hat{1},\hat{I})$ is the group consisting of the identity $\hat{1}$ and the inversion operator $\hat{I}$ of a coordinate through the mid-point of the internuclear axis (see sec. V.1), $\mathscr{C}_{\infty v}$ the group containing operators $\hat{C}(\phi)$ of rotation by any angle ϕ about the internuclear axis, an operator $\hat{\sigma}$ for reflection through the mirror plane containing the internuclear axis and operator products $\hat{C}(\phi)\hat{\sigma}$, while $S_2 \equiv (\hat{1},\hat{12})$ is the symmetric group of order two consisting of the identity and the operator $\hat{12}$ which permutes the electron coordinates between each other. In general, the group to be used in symmetry-adapted perturbation theories can be reduced with respect to the full group G because, loosely speaking,

Table VIII.1. Interaction energy values (a.u.) for the states $X^1\Sigma_g^+$ and $b^3\Sigma_u^+$ of H_2, as a function of the internuclear distance R (a.u.) from several perturbation theory procedures [313]. [a(-n) ≡ a·10^{-n}].

Method	R = 4.0 a.u.	R = 6.0 a.u.	R = 8.0 a.u.	R = 10.0 a.u.
		$X^1\Sigma_g^+$		
HL	-1.1286(-2)	-0.5092(-3)	-0.1740(-4)	-0.5153(-6)
P-HL	-1.3455(-2)	-0.7155(-3)	-0.5037(-4)	-0.8327(-5)
MS-MA	-1.4080(-2)	-0.7497(-3)	-0.5194(-4)	-0.8395(-5)
EL-HAV	-1.4262(-2)	-0.7316(-3)	-0.4636(-4)	-0.6854(-5)
PA	-0.3778(-2)	-0.2660(-3)	-0.3476(-4)	-0.7860(-5)
CI	-1.5836(-2)	-0.7950(-3)	-0.5225(-4)	-0.8354(-5)
Accurate	-1.6369(-2) [a]	-0.8342(-3) [b]	-0.5544(-4) [b]	-0.8738(-5) [b]
		$b^3\Sigma_u^+$		
HL	0.8790(-2)	0.3919(-3)	0.1382(-4)	0.4187(-6)
P-HL	0.6620(-2)	0.1856(-3)	-0.1915(-4)	-0.7394(-5)
MS-MA	0.7518(-2)	0.2203(-3)	-0.1758(-4)	-0.7326(-5)
EL-HAV	0.7193(-2)	0.2382(-3)	-0.1201(-4)	-0.5785(-5)
PA	-0.3778(-2)	-0.2659(-3)	-0.3476(-4)	-0.7860(-5)
CI	0.6739(-2)	0.2054(-3)	-0.1766(-4)	-0.7296(-5)
Accurate	0.6622(-2) [a]	0.0019(-5) [b]	-0.2015(-4) [b]	-0.7663(-5) [b]

(a) Ref. [313]; (b) Ref. [66].

a part of the symmetry properties of the eigenstates to $\hat{H}$ are already contained in the unperturbed eigenstate φ_0 to $\hat{H}_0$. Thus, if we observe that $\mathcal{C}_{\infty v}$ commutes with both $\hat{H}$ and $\hat{H}_0$ and that $\varphi_0(1,2) \equiv 1s_a(1)1s_b(2)$ belongs to the Σ^+ irreducible representation of $\mathcal{C}_{\infty v}$, we can get rid of $\mathcal{C}_{\infty v}$. Moreover, the subgroup $\mathcal{C}_i$ is isomorphic to S_2: any symmetry-adapted perturbation theory for the interaction H···H can therefore be founded <u>exclusively</u> on the explicit consideration of the <u>symmetric</u>

group S_2 and its projection operators $\hat{Q}^{(+)} \equiv \rho^{[2]} = \frac{1}{2}[\hat{1} + (\hat{12})]$, $\hat{Q}^{(-)} \equiv \rho^{[1^2]} = \frac{1}{2}[\hat{1} - (\hat{12})]$, corresponding to singlet and triplet states, respectively (see Chap.2). (A formal proof of the result presented here on an intuitive basis is given in ref. [313]).

Values of the interaction energy for the lowest states of singlet ($X^1\Sigma_g^+$) and triplet ($b^3\Sigma_u^+$) symmetry of two interacting hydrogen atoms are reported in Table VIII.1. With the exception of the quantity labeled HL, which is simply the 1-st order Heitler-Landon energy, the energies labelled P-HL, MS-MA, EL-HAV and PA correspond to values inclusive of 2-nd order contributions (see also Table V.1). P-HL, the sum of the 1-st order HL energy and the 2-nd order polarization energy (see Table V.2) differs from PA, the energy through the 2-nd order as evaluated in the " polarization approximation", in the exchange contributions present in the 1-st order HL energy. The various 2-nd order perturbation results collected in the Table, even though not exact (there are not rigorously exact results at disposal for the case under investigation), should be regarded as good approximations to the true energies through the 2-nd order. They have been obtained, in fact, by approximating the relevant 1-st order perturbed wavefunction by series expansion in terms of an extended basis set of Slater-type orbitals, after reducing the solution of the involved differential equation to the search of the stationary points of a proper Hylleraas-type functional [37-39] (see sec. V.1). An appreciation of the quality of the results obtained can be gained by comparing the predictions from the various perturbation theories with the values labelled CI, derived by a configuration interaction procedure in terms of the full number of configurations arising from the same basis set employed in the perturbative approach. The difference between CI and accurate values, in fact, can be filled only by enlarging the subspace spanned by the basis set used.

As a brief comment to the results collected in Table VIII.1, we observe that they give further support to the conclusions drawn from the inspection of Table V.1 relative to the interaction $H\cdots H^+$. So, the inadequacy of the 1-st order HL energy and the consequent necessity of pushing the perturbation procedure to the 2-nd order come out once more, along with the eminent reasonableness of the MS-MA approach at intermediate and large distances. The EL-HAV theory, as expected, displays its insufficient behavior at large R values, where the "polarization approximation" PA tends to become adequate (after having been inadequate in the other regions), whereas P-HL, which simply combines the 1-st order HL energy with the 2-nd order polarization energy, appears a rather considerable approximation, which legitimates its application, to more complicate interactions than H•••H.

As far as the $b^3\Sigma_u^+$ state is concerned, one notes even a very remarkable agreement between P-HL and accurate results, probably a rather fortuitous occurrence caused by favourable cancellation of missing interaction contributions.

The relative importance of induction and dispersion contributions to the 2-nd order polarization energy does not emerge from the data of Table VIII.1, but can be appreciated from the Tables VII.7 and VII.8 (which, incidentally, confirm the values for the 2-nd order polarization energies used in Table VIII.1).

b) The determination of the interaction energy between rare gas atoms has received much attention particularly from the experimental point of view (for a fairly recent, very good survey of approaches founded on the utilization of thermodynamic and non-thermodynamic data, see for example ref. [5]). The computational effort, comparably less intense, shows a prevalence of variational calculations with respect to perturbation theory applications, which are essentially restricted to the case of the interaction He•••He between two ground state helium atoms [40,41,157,319-322]. As far as we are concerned, instead of stressing the application of perturbation theories to such interaction, we prefer to emphasize the results of a <u>simplified</u> P-HL procedure which has recently been suggested and applied to situations involving heavier rare gas atoms [322]. The most drastic changes brought to the P-HL scheme consist in i) approximating the <u>whole</u> 2-nd order energy contribution over the entire range investigated by its dispersion component in <u>multipolar</u> form, i.e. neglecting charge overlap effects (see sec. VII.4), and ii) evaluating the interaction energy with respect to a "zero-th order" associated with atomic basis functions which are not eigenstates to the separate Hamiltonian operators (that requires the explicit introduction of the "complementary exchange" term Δ considered in sec. VI.6). Other minor approximations involve the neglect of some contributions to the interaction energy arising from orbitals which have a very small interatomic overlap (inner shell orbitals, for example).

The interaction energy $\Delta W(R)$ for a pair of rare gas atoms at distance R is written in the form

$$\begin{aligned}\Delta W(R) &\equiv W(R) - (\langle\phi_0^A|H_0^A|\phi_0^A\rangle + \langle\phi_0^B|\hat{H}_0^B|\phi_0^B\rangle) \\ &\simeq W_{01} + W_{01,ex} + \Delta - \sum_{n=3}^{5} C_{2n}^{disp} R^{-2n}\end{aligned} \qquad (VIII.1)$$

where $W_{01} + W_{01,ex}$ represents the (approximate) 1-st order HL energy [see eq. (V.3.8)] and C_6^{disp}, C_8^{disp}, C_{10}^{disp} are van der Waals dispersion energy coefficients corresponding to dipole-dipole, dipole-quadrupole and (dipole-octopole + quadrupole-quadrupole) interactions, respectively. In spite of the simplicity of the model, "result as accurate as those obtained by the variational method in the region of the van der Waals minimum and perhaps more accurate at larger internuclear separations" are claimed to be within its range [322] provided that atomic wavefunctions $|\phi_0^A\rangle$ and $|\phi_0^B\rangle$ close to the HF limit are employed; the rather dubious accuracy of the higher multipole dispersion coefficients C_8^{disp}, C_{10}^{disp} for all the interactions except He•••He, which appears a not completely reassuring matter, does not seem to be too detrimental.

Curves representing the interaction energy $\Delta W(R)$ between two ground state neon atoms and two ground state argon atoms as a function of the separation R are drawn in Figs. VIII.1 and VIII.2, respectively. The behaviour of $\Delta W(R)$ as evaluated according to the simplified P-HL approach just examined is contrasted with that from different procedures. For Ne•••Ne, calculated and experimental potentials are close to each other [there is also confirmation for a modest role of the exchange contributions arising from the (neglected) orbitals with small intermolecular overlap], while the deviation between calculation and experiment

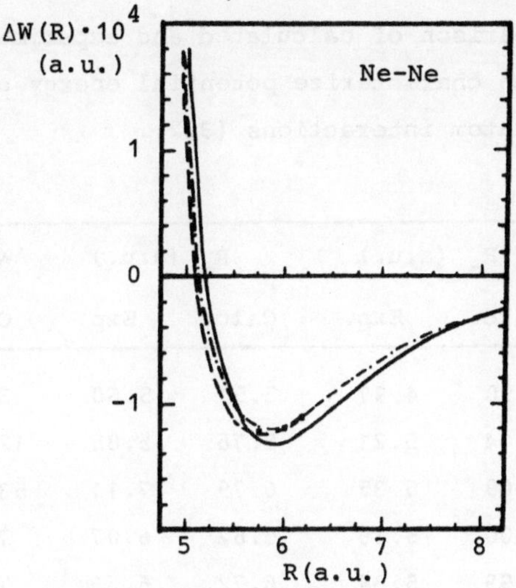

Fig. VIII.1. Potential energy curves for the interaction Ne···Ne: a comparison among simplified P-HL approach (eq.VIII.1) (dashed) and other source calculations (dash-dot: ref.[321]; experimental potential [323]: full line. The points represent SCF CI calculations [324]).

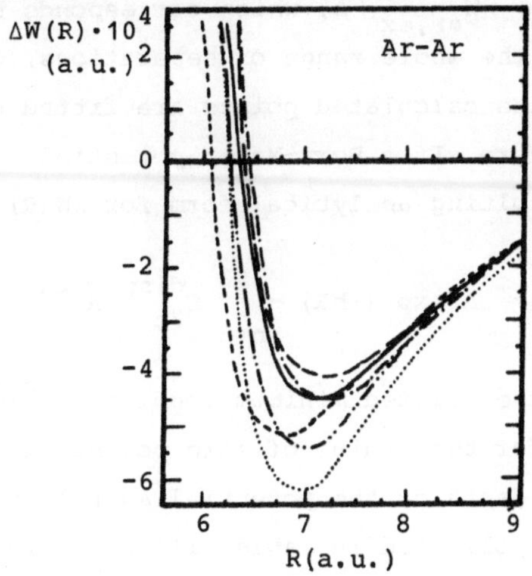

Fig. VIII.2. Potential energy curves for the interaction Ar···Ar. — — — : ref. [325]; – – –: ref. [326]; —·—·— : ref. [327]; —··—··— : ref.[328]; ···· : ref. [329]; ----- (simplified P-HL approach): ref. [322].

Table VII.2. Comparison of calculated and experimental parameters which characterize potential energy curves for rare gas atom interactions [322].

	R_o (a.u.)		R_m (a.u.)		$\Delta W(R_m)\cdot 10^5$ (a.u.)	
	Calc.	Exp.	Calc.	Exp.	Calc.	Exp.
He···He	4.90	4.97	5.51	5.60	3.77	3.49
Ne···Ne	5.14	5.21	5.76	5.88	12.67	12.89
Ar···Ar	6.09	6.35	6.79	7.11	53.95	45.00
He···Ne	5.00	5.16	5.62	6.07	7.09	4.53
He···Ar	5.99	5.84	6.72	6.69	7.78	7.66
Ne···Ar	5.90	5.84	6.62	6.48	17.69	22.8

is larger for the case Ar···Ar. The 1-st order contribution to the interaction energy, $W_{01} + W_{01,ex} + \Delta$, which corresponds to a globally repulsive effect over the whole range of separations, can be expressed in a useful way if the calculated points are fitted to some proper analytical functional form. If a Born-Mayer potential function A exp(-bR) is employed, the resulting analytical form for $\Delta W(R)$ becomes

$$\Delta W(R) = A \exp(-bR) - \sum_{n=3}^{5} C_{2n}^{disp} R^{-2n} \qquad \text{(VIII.2)}$$

If eq.(VIII.2) and the related minimum condition $[\frac{d\Delta W(R)}{dR}]_{R=Rm} = 0$ are numerically solved for the radius of zero potential R_0, the equilibrium distance R_m and the depth of the potential $\Delta W(R_m)$ at its minimum, one obtains the results collected in Table VIII.2, where comparisons with experimental estimates are also presented. A good agreement between calculations and experiments is evident for the interactions He···He, Ne···Ne and He···Ar, along with discrepancies for Ar···Ar, He···Ne and Ne···Ar. It is claimed, however, that the experimental curves for the mixed interactions might be less firmly established than for homonuclear diatomics [322].[†]

c) Perturbative calculations of interaction energies between molecules are scarce, a fact not surprising in view of the nature of the problem, that combines the basic complexity of a many-electron system with the complications arising from the presence of several centres and its anisotropic character. At the time of ending these notes, we are aware of molecular perturbative calculations concerning only the interactions $H_2\cdots H_2$ between two ground state hydrogen molecules [330-32] and $H_2O\cdots H_2O$, between two ground state water molecules [308,333]. In view of the important role played by the latter interaction in physical chemistry and biophysics (hydrogen bonding), we think it suitable to terminate our survey of examples dwelling upon the interaction $H_2O\cdots H_2O$ on the basis of results obtained in a recent and valuable work [308].

When passing to deal with interactions between many-electron "monomers" of appreciable complexity, in addition to unavoidable approximations connected with the more or less complete neglect of electron correlation effects, it becomes quite understandable any attempt aimed to reduce the size of the computational burden, providing that the further approximations introduced do not distort the essence of the original procedure. On the grounds of the results relative to the simple situations examined in the preceding pages and bona fide extrapolations to more complex problems, the P-HL approximation can be judged a reasonably acceptable compromise between (sufficient) accuracy and simplicity (it affords the part in common that all the symmetry-adapted perturbation theories share with each other).

The calculations to which we shall make reference are founded exclusively on SCF wavefunctions of high quality (not too far from the

---

(†) We point out that, contrary to the discussion of the interaction $H\cdots H$, where a spin-free point of view was adopted, a symmetry-adapted perturbation theory founded on wavefunctions depending on space and spin variables has been preferred in the present case, thus making the formalism developed in Chap.V applicable in a direct way. A completely equivalent spin-free procedure, entirely based on the use of Young operators $\hat{\rho}^{[\tau]}$ acting on purely spatial eigenstates (see Chap.II), is obviously possible, but has found only rare implementation in many-electron problems [see for example, ref. [307])].

H.F. limit), so they neglect completely correlation effects. This is surely a serious drawback, because the 2-nd order dispersion energy is known to be among the various contributions rather sensitive to the "intramonomer" correlation corrections, and the same is true, even though possibly to a smaller extent, for the 1-st order exchange energy (see preceding Chapters). In spite of this, we are faced with an important result, because if provides us with a reliable value for the

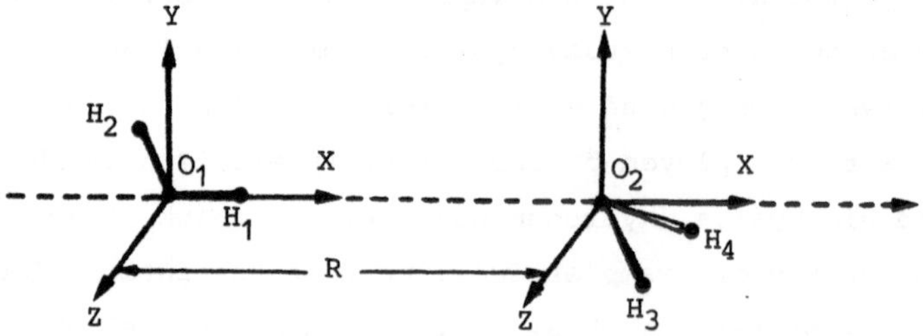

Fig. VIII.3. $H_2O\cdots H_2O$ interaction: relative (fixed) orientation of the two molecules considered in the calculations (linear dimer) [308].

leading term of the series expansion of the P-HL energy in terms of "intramonomer" correlation corrections (see sec. V.4). All numerical calculations have been performed for a fixed orientation of the two molecules (see Fig. VIII.3) and several values of the separation R between the two oxygen nuclei O_1, O_2. The evaluated SCF energy of the water monomer, -76.0576 a.u., is only 0.11 % far from the recent <u>Popkie et al</u>. SCF value, -76.0660 a.u., which is considered very near the H.F. limit [334]. The various contributions to the P-HL energy through the 2-nd order ΔW are collected in Table VIII.3, for values of the separation R between the two oxygen nuclei in the range 4.00 ÷ 15.00 a.u.. The value $R = 5.67$ a.u. corresponds to the minimum of the energy of the dimer in the configuration of Fig. VIII.1 as obtained by a SCF calculation in terms of the same basis set employed in the perturbative calculation here considered. The important role of the 1-st order attractive and repulsive contributions to the interaction energy ΔW are immediately recognized, along with the necessity of taking into account 2-nd order contributions in the zone of the SCF minimum R = 5.67 a.u.. Here "the high polarizability of the water molecules becomes decisive and causes the dispersion and induction energies to reach appreciable values and in consequence to stabilize the hydrogen bonding" [308]. One also observes that at smaller R values the induction energy is more important than the dispersion one, while at large distances the contrary is the case.

Table VIII.3. Contributions to the interaction energy ΔW of two H_2O molecules (in Kcal/mole) as a function of the separation R_{O-O} at fixed relative orientation (see Fig. VIII.1) [308].

R(a.u.)	W_{01}	$W_{01,ex}$	$W_{02,ind}$	$W_{02,disp}$	ΔW	ΔW_{CI} [a]
4.00	-43.09	105.49	-49.30	-14.46	-1.36	
4.40	-27.10	51.58	-21.00	-8.41	-4.93	
4.50						+5.67
4.80	-17.10	25.11	-9.30	-4.89	-6.15	
5.00						-3.58
5.20	-11.10	12.03	-4.12	-2.85	-6.04	
5.50						-5.55
5.67	- 7.12	4.90	-1.63	-1.54	-5.39	
6.00						-5.32
7.00	- 2.79	0.30	-0.18	-0.31	-2.98	-3.04
9.00	- 1.12	0.00	-0.02	-0.05	-1.19	
15.00	- 0.21	0.00	-0.00	-0.00	-0.21	

a) Ref. [335].

The entries of Table VIII.3 show clearly the important role of the dispersion energy, which not only increases (in absolute value) the dimerization energy but also shifts the energy minimum towards smaller

R distances. The values of $W_{02,disp}$ are the same obtainable by using the rather rough uncoupled H.F. perturbation method reviewed in Chap. VII, which is known to lead to generally underestimated polarizabilities and dispersion energy coefficients as well. An increased depth of the dimerization energy, which could be expected on the basis of this insufficient appreciation of the dispersion energy contribution, is not taken for granted, however, because neglected 2-nd order exchange term and intracorrelation correction to the 1-st order energy work likely in the opposite direction. A reliable experimental value for the dimerization energy of H_2O is not available, even though probably it lies between 4.8 and 6.5 Kcal/mole, as deduced from Monte Carlo calculations carried out on liquid water [336]. The value 6.15 Kcal/mole in Table VIII.3 is within such range, and compares favourably with an independent estimate [335] obtained by an extensive configuration interaction calculation (seventh column in Table VIII.3). The comparison between ΔW and ΔW_{CI} values, however, reveals that the matter is not completely settled.

APPENDIX A

The theoretical framework reviewed in Chap. III affords a general basis for generating symmetry - adapted perturbation theories. Through the use of the partitioning technique we have actually obtained some current "exchange" perturbation procedures, which were originally derived in rather different ways. An unified approach to various "exchange" theories has already been pursued by several people [37,337-341]: since we think that some further insight into the nature of these theories can be gained by looking upon them from different standpoints, we shall now present one of these analyses along the lines suggested by Chipman [37,340].

The basic problem we are faced with involves the solution of the N-electron Schrödinger equation for the possible states of the system constituted by a given number of interacting subsystems

$$\hat{H} \mid \Psi_r^{[\tau]}\rangle = E_r^{[\tau]} \mid \Psi_r^{[\tau]}\rangle \tag{A.1}$$

We are considering a <u>spin-free</u> approach [14,341,343], so that $|\Psi_r^{[\tau]}\rangle$ refers to the <u>space</u> part of the electronic eigenstate, which transforms according to the irreducible representation $[\tau]$ of the symmetric group S_N, i.e. $\hat{\rho}^{[\tau]}|\Psi^{[\tau]}\rangle = |\Psi^{[\tau]}\rangle$, $\hat{\rho}^{[\tau]}$ being the proper projection operator, eq. (II.2.5); we are also assuming that no further degeneracies, apart from that associated with S_N, an involved.

Our program is to find a solution to eq. (A.1) starting from an exactly known solution to the unperturbed problem

$$\hat{H}_0 \mid \varphi_0\rangle = E_0 \mid \varphi_0\rangle \tag{A.2}$$

the perturbation being $\hat{V} = \hat{H} - \hat{H}_0$. To this aim, let us consider the (inhomogeneous) equation

$$[\hat{H}_0 + \lambda\hat{V}' - \varepsilon^{[\tau]}(\lambda)]\,|\phi(\tau;\lambda)\rangle = (\hat{1}-\hat{\rho}^{[\tau]})\,|F(\tau;\lambda)\rangle \tag{A.3}$$

where λ is a continuous parameter and $|F(\tau;\lambda)\rangle$ an essentially <u>arbitrary</u> vector, that we assume to be analytic in λ and such that $|F(\tau;o)\rangle \equiv 0$. Thanks to this latter requirement, eq. (A.3) reduces to the unperturbed

Schrödinger equation (A.2) as $\lambda \to 0$, so that the λ-expansions for $|\phi(\tau;\lambda)>$ and $\varepsilon^{[\tau]}(\lambda)$ will be of the form

$$|\phi(\tau;\lambda)> = |\varphi_0> + \lambda|\phi_1(\tau)> + \lambda^2|\phi_2(\tau)> + \dots$$
$$\varepsilon^{[\tau]}(\lambda) = E_0 + \lambda E_{01}^{[\tau]} + \lambda^2 E_{02}^{[\tau]} + \dots \qquad \text{(A.4)}$$

Once a solution $|\phi(\tau;\lambda)>$ to eq. (A.3) has been obtained, there are various ways for deriving a solution to the original Schrödinger equation (A.1). If $|F(\tau;\lambda)>$ is chosen in such a way that $(\hat{1} - \hat{\rho}^{[\tau]})|F(\tau;1)> = 0$, then at $\lambda = 1$ eq. (A.3) becomes identical to eq. (A.1) and (provided that $|\phi(\tau;1)> \neq 0$) a solution to the Schrödinger equation is given by $|\psi_r^{[\beta]}> \equiv |\phi(\tau;1)>$, $E_r^{[\beta]} = \varepsilon^{[\tau]}(1)$, for same $[\beta]$ and r. It should be remarked that, contrary to $|\phi(\tau;1)>$, a perturbation expansion of $|\phi(\tau;\lambda)>$ truncated after a <u>finite</u> order does not necessarily have the symmetry of an exact eigenstate $|\psi_r^{[\beta]}>$ to $\hat{H}$. A second general way for recovering a solution to the Schrödinger equation from $|\phi(\tau;\lambda)>$ is easily found on applying the projector $\hat{\rho}^{[\tau]}$ to both sides of eq. (A.3); the result of such operation is

$$\hat{\rho}^{[\tau]}[\hat{H}_0 + \lambda\hat{V} - \varepsilon^{[\tau]}(\lambda)]|\phi(\tau;\lambda)> = 0 \ .$$

Now, as $\lambda = 1$, $[\hat{\rho}^{[\tau]}, \hat{H}_0 + \hat{V}] = 0$, and therefore

$$[\hat{H}_0 + \hat{V} - \varepsilon^{[\tau]}(\lambda)]\hat{\rho}^{[\tau]}|\phi(\tau;\lambda)> = 0 \ ,$$

so that (provided that $\hat{\rho}^{[\tau]}|\phi(\tau;1)> \neq 0$) we get $|\psi_r^{[\tau]}> = \hat{\rho}^{[\tau]}|\phi(\tau;1)>$, $E_r^{[\tau]} = \varepsilon^{[\tau]}(1)$. This second method is characterized by the fact that an eigenstate of the correct symmetry is obtained even though the perturbative series for $|\phi(\tau;\lambda)>$ is truncated at a finite order.

For a given choice of $|F(\tau;\lambda)>$ we have obviously a well definite equation (A.3), hence a given set of perturbation equations for the various order corrections $|\phi_{on}(\tau)>$ to $|\varphi_0>$ and consequently a perturbation series conceivably endowed with peculiar convergence properties, possibly quite different from those of the series arising from another choice of the inhomogeneity. From the present standpoint, therefore, the various possible "exchange" perturbation theories correspond to different choices of the inhomogeneity $|F(\tau;\eta)>$ in eq. (A.3). We shall now examine some special cases of the general scheme.

a) Polarization Approximation PA [5,6,55].

This actually non-symmetry-adapted procedure follows from eq. (A.3) by the choice $|F(\tau;\lambda)> \equiv 0$. In this way eq. (A.3) transforms into the homogeneous Schrödinger equation

$$[\hat{H}_0 + \lambda\hat{V} - \varepsilon^{(PA)}(\lambda)]|\phi^{(PA)}(\lambda)> = 0 \qquad (A.5)$$

the starting point for the ordinary Rayleigh-Schrödinger or Brillouin-Wigner perturbation theory. It should be observed how any reference to the action of the projector $\hat{\rho}^{[\tau]}$ has become lost, so that the label $[\tau]$ has been dropped: we have, therefore, one and the same eq. (A.5) for all possible $[\tau]$'s. Apart from convergence problems, an exact solution to eq. (A.1) is recovered according to $|\psi_r^{[\tau]}> = |\phi^{(PA)}(1)>$, $E_r^{[\tau]} = \varepsilon^{(PA)}(1)$, for some $[\tau]$ and r. Since $[\tau]$ denotes one of the irreducible representations of S_N, it is clearly possible that the procedure leads to Pauli excluded representations (see Chap. II), as explicitly shown by the analyses presented in sec. IV.3.

b) Hirschfelder-Silbey (HS) method [43,344].

This peculiar symmetry-adapted perturbation theory can be generated from eq. (A.3) by choicing $|F(\tau;\lambda)> = [-\varepsilon^{[\tau]}(\lambda) + \sum_{[\tau']} \varepsilon^{[\tau']}(\lambda)\hat{\rho}^{[\tau']}]|\phi(\tau;\lambda)>$
As one easily verifies, the inhomogeneous equation (A.3) is transformed into the homogeneous one,

$$[\hat{H}_0 + \lambda\hat{V} - \sum_{[\tau']} \varepsilon^{[\tau']}_{(HS)}(\lambda)\ \hat{\rho}^{[\tau']}]|\phi^{(HS)}(\lambda)> = 0 \qquad (A.6)$$

the $[\tau']$ sum being extended to all the irreducible representations of S_N. The same eq. (A.6) is obtained, regardless of the choice of $[\tau]$, so that the label $[\tau]$ has been dropped away. In order to recover the solution to the original Schrödinger equation, we let operate $\rho^{[\tau]}$ on both sides of eq. (A.6); then, at $\lambda = 1$, we find $|\psi_r^{[\tau]}> = \hat{\rho}^{[\tau]}|\phi^{(HS)}(1)>$,

$E_r^{[\tau]} = \varepsilon_{(HS)}^{[\tau]}(1)$. It should be clear that in general all the s projectors $\hat{\rho}^{[\tau]}$ will give results different from zero where operating on $|\phi^{(HS)}(1)>$, so that a single "primitive" eigenvector $|\phi^{(HS)}>$ provides a number of solutions, including those Pauli excluded, to the Schrödinger equation.

Eq. (A.6) does not determine the single energies $\varepsilon_{(HS)}^{[\tau]}$; to this end, we consider the original Schrödinger equation (A.1), from which we get $<\varphi_0|\hat{V} - (E_r^{[\tau]} - E_o)|\psi^{[\tau]}> = 0$; but $|\psi_r^{[\tau]}> = \hat{\rho}^{[\tau]}|\phi^{(HS)}(1)>$, so that

$$<\varphi_0|\hat{V}\hat{\rho}^{[\tau]}|\phi^{(HS)}(1)> = (E_r^{[\tau]} - E_0)<\varphi_0|\hat{\rho}^{[\tau]}|\phi^{(HS)}(1)> = \tag{A.7}$$

$$= [\varepsilon_{(HS)}^{[\tau]}(1) - E_0]<\varphi_0|\hat{\rho}^{[\tau]}|\phi^{(HS)}(1)>$$

which provides a relation for $\varepsilon_{(HS)}^{[\tau]}(1)$.

If a perturbative treatment is applied to eq. (A.6), the correct eigenstate through the 1-st order results $|\varphi_0> + \lambda|\phi_{01}^{(HS)}>$, where $|\phi_{01}^{(HS)}> = \hat{R}_0[\hat{V} - \sum_{[\tau']} E_{01(HS)}^{[\tau']}\hat{\rho}^{[\tau']}]|\varphi_0>$ (intermediate normalization $<\varphi_0|\phi^{(HS)}(\lambda)> = <\varphi_0|\varphi_0> = 1$ has been assumed). From eq. (A.7) we then obtain the following few corrections $E_{on(HS)}^{[\tau]}$ to E :

$$E_{01(HS)}^{[\tau]} = \frac{<\varphi_0|\hat{V}\hat{\rho}^{[\tau]}|\varphi_o>}{<\varphi_o|\hat{\rho}^{[\tau]}|\varphi_o>}$$

$$E_{02(HS)}^{[\tau]} = \frac{<\varphi_0|[\hat{V} - E_{01(HS)}^{[\tau]}]\hat{\rho}^{[\tau]}|\phi_{01}^{(HS)}>}{<\varphi_o|\hat{\rho}^{[\tau]}|\varphi_o>} \tag{A.8}$$

We recognize that $E_{01(HS)}^{[\tau]}$ has the standard form found from any symmetry-adapted perturbation procedure (1-st order Heitler-London energy), while $E_{02(HS)}^{[\tau]}$ is characterized by a peculiar structure depending on $|\phi_{01}^{(HS)}>$ and different from E_{02} as evaluated in other schemes (MS-MA, EL-HAV, etc.).

c) <u>Murrell-Shaw, Musher-Amos (MS-MA) method</u> [36-39,45].

This perturbation approach to the "exchange" problem corresponds, in the general scheme, to the choice $|F(\tau;\lambda)\rangle = [\hat{H}_0 + \lambda\hat{V} - \varepsilon^{[\tau]}(\lambda)]|\varphi_0\rangle$ for the inhomogenity in eq. (A.3). It is then a simple exercise to verify that eq. (A.3) becomes

$$[\hat{H}_0 + \lambda\hat{V} - \varepsilon^{[\tau]}_{(MSMA)}(\lambda)][|\phi^{(MSMA)}(\tau;\lambda)\rangle - (\hat{1} - \hat{\rho}^{[\tau]})|\varphi_o\rangle] = [\hat{H}_o + \lambda\hat{V}, \hat{\rho}^{[\tau]}]|\varphi_o\rangle \tag{A.9}$$

where at the right-hand side appears the commutator of $\hat{H}_0 + \lambda\hat{V}$ with $\hat{\rho}^{[\tau]}$.

A solution to the original Schrödinger equation (A.1) can easily be recovered from that of eq. (A.9) if we observe that at $\lambda = 1$ the right-hand side of eq. (A.9) vanishes identically (because $\hat{\rho}^{[\tau]}$ commutes with the Hamiltonian operator $\hat{H}_0 + \hat{V}$). Hence a solution to eq. (A.1) is given by

$$|\psi_r^{[\tau']}\rangle = |\phi^{(MSMA)}(\tau;1)\rangle - (1 - \rho^{[\tau]})|\varphi_o\rangle$$

$$E_r^{[\tau']} = \varepsilon^{[\tau]}_{(MSMA)}(1) \tag{A.10}$$

for some $[\tau']$ and r.

From eq. (A.10) it is observed that, if the right-hand side of the 1-st equation is truncated after the zero-th order, the resulting approximate state vector has the correct permutational symmetry $[\tau]$; the same, however, is not generally true if the truncation occurs at any other finite order. These remarks are manifestly in line with those of sec. III.4 (see the last paragraph of that section).

A perturbative solution of eq. (A.9) leads in a natural way to

$$E^{[\tau]}_{01\,(MSMA)} = \frac{\langle\varphi_0|\hat{\rho}^{[\tau]}\hat{V}|\varphi_o\rangle}{\langle\varphi_o|\hat{\rho}^{[\tau]}|\varphi_o\rangle}$$

$$E^{[\tau]}_{o2\,(MSMA)} = \frac{\langle\varphi_o|\hat{V}|\phi^{(MSMA)}_{o1}\rangle}{\langle\varphi_o|\hat{\rho}^{[\tau]}|\varphi_o\rangle} \tag{A.11}$$

where $|\phi_{01}^{(MSMA)}\rangle = \hat{R}_0\hat{\rho}^{[\tau]}(\hat{V} - E_{01(MSMA)}^{[\tau]})|\varphi_0\rangle$, the same result already obtained in sec. III.4 through the partitioning technique.

d) Eisenschitz-London, Hirschfelder-van der Avoird (EL-HAV) method [42-45].

The EL-HAV method follows from eq. (A.3) by making the choice

$$F(\tau;\lambda) = [\lambda\hat{V} - \varepsilon^{[\tau]}(\lambda) + E_0]|\phi(\tau;\lambda)\rangle \tag{A.12}$$

for the inhomogeneity. It is immediately seen that we find in this case

$$[(\hat{H}_0 - E_0) + \hat{\rho}^{[\tau]}(\lambda\hat{V} - \varepsilon_{(ELHAV)}^{[\tau]}(\lambda) + E_0)]|\phi^{(ELHAV)}(\tau;\lambda)\rangle = 0 \tag{A.13}$$

from the solution of which equation we can recover a solution to the original Schrödinger equation (A.1) by projection, i.e.

$$\begin{aligned} |\psi_r^{[\tau]}\rangle &= \hat{\rho}^{[\tau]}|\phi^{(ELHAV)}(\tau;1)\rangle \\ E_r^{[\tau]} &= \varepsilon_{(ELHAV)}^{[\tau]}(1) \end{aligned} \tag{A.14}$$

It is then a straightforward exercise to obtain the first few corrections $E_{0n(ELHAV)}$ to the energy E_0 by perturbatively solving eq. (A.13).

APPENDIX B

In this Appendix we propose to analyze in some detail the long-range behaviour of the second-order energy for H_2^+ as predicted by different symmetry-adapted perturbation theories.

We start by considering the "polarization approximation", because the result we shall obtain is entirely transferable to MS-MA. Making use of the multipole expansion for $\hat{V}$, according to the lines put forward in sec. V.2, we get

$$E_{02}^{as} = \sum_{\ell=1}^{\infty} E_{02}^{(\ell)as} = \sum_{\ell=1}^{\infty} \langle \varphi_{01}^{(\ell)as} | \hat{V}^{as} | \varphi_0 \rangle \tag{B.1}$$

The asymptotic contribution $\hat{V}^{as}$ to the perturbation potential energy $\hat{V}$, introduced in the same section above quoted, is

$$\hat{V}^{as} = \sum_{\ell=1}^{\infty} V^{as}(r_a, R)\, P_\ell(\cos\theta_a) \tag{B.2}$$

$$\hat{V}_\ell^{as}(r_a, R) = -r_a^\ell / R^{\ell+1} \tag{B.3}$$

$|\varphi_{01}^{as}\rangle$ is the long-range solution to the 1-st order perturbation equation

$$(\hat{H} - E_0)|\varphi_{01}\rangle = (E_{01} - \hat{V})|\varphi_0\rangle$$

[see eq.(V.1.3)], i.e. the equation which descends from the previous one by replacing $\hat{V}^{as} \rightarrow \hat{V}$, $E_{01} = \langle\varphi_0|\hat{V}|\varphi_0\rangle \rightarrow 0$ as a consequence of (B.2) and the spherical symmetry of $|\varphi_0\rangle$ around the center a. If we make $\hat{H}_0$ explicit, the differential equation to be solved results

$$[-\frac{1}{2}\nabla^2 - \frac{1}{r_a} + \frac{1}{2}]\ \varphi_{01}^{as}(r_a, \theta_a) = \sum_{\ell=1}^{\infty} \frac{r_a^\ell}{R^{\ell+1}} P_\ell(\cos\theta_a) 1s_a(r_a) \tag{B.4}$$

As a first step in the solution of eq. (B.4) we put

$$\varphi_{01}(r_a, \theta_a) = \sum_{\ell=1}^{\infty} \varphi_{01}^{(\ell)as}(r_a, \theta_a) = \sum_{\ell=1}^{\infty} \frac{f_\ell(r_a)}{R^{\ell+1}} P_\ell(\cos\theta_a) \tag{B.5}$$

into eq. (B.4), so that the following equation for the amplitude $f_\ell(r_a)$ results:

$$\frac{1}{2r_a}\frac{d^2}{dr_a^2}(r_a f_\ell) - \left[\frac{\ell(\ell+1)}{2r_a^2} - \frac{1}{r_a} + \frac{1}{2}\right] f_\ell = -r_a^\ell \, \pi^{-\frac{1}{2}} e^{-r_a}$$

If we then search for a solution of the form

$$f_\ell(r_a) = 2\pi^{-\frac{1}{2}} e^{-r_a} g_\ell(r_a) \tag{B.6}$$

we find readily that $g_\ell(r_a)$ must satisfy the inhomogeneous equation

$$\frac{d^2 g_\ell}{dr_a^2} + 2\left(\frac{1}{r_a} - 1\right)\frac{dg_\ell}{dr_a} - \frac{\ell(\ell+1)}{r_a^2} g_\ell = -r_a^\ell \tag{B.7}$$

whose solution is easily verified to be

$$g_\ell(r_a) = \frac{1}{2}\left(\frac{r_a^\ell}{\ell} + \frac{r_a^{\ell+1}}{\ell+1}\right) \tag{B.8}$$

The long-range 2-nd order energy E_{02} can now be evaluated in a straightforward way. From eqs. (B.1), (B.2), (B.3) and (B.5), taking into account the orthogonality of the Legendre polynomials, it follows

$$E_{02}^{as}(R) = -4\pi \sum_{\ell=1}^{\infty} (2\ell+1)^{-1} R^{-2(\ell+1)} \int_0^\infty dr_a \, r_a^{\ell+2} \, f_\ell(r_a)\varphi_0(r_a)$$

and successively, from eqs. (B.6), (B.8)

$$E_{02}^{as}(R) = -4 \sum_{\ell=1}^{\infty} (2\ell+1)^{-1} R^{-2(\ell+1)} \int_0^\infty dr_a \, e^{-2r_a}\left[\frac{r_a^{2\ell+2}}{\ell} + \frac{r_a^{2\ell+3}}{\ell+1}\right]$$

After some simple manipulations we find the final result

$$E_{02}^{as}(R) = -\sum_{\ell=1}^{\infty} \frac{(2\ell+2)!\,(\ell+2)}{2^{2\ell+2}\,\ell(\ell+1)} \cdot R^{-2(\ell+1)} \tag{B.9}$$

that we have quoted as eq. (V.2.11).

As far as the long-range behaviour of the 2-nd order energy $E_{02}^{(i)}$ from EL-HAV is concerned, we start from eq. (III.4.13) which provides a relation between $E_{02}^{(i)}$ evaluated from EL-HAV and $E_{02}^{(i)}$ from MS-MA:

$$E_{02}^{(i)}(\text{EL-HAV}) = E_{02}^{(i)}(\text{MS-MA}) - N^{-1}\langle\varphi_{01}|\hat{\omega}_i^{+}[\hat{\rho}^{(i)},\hat{V}]\hat{\omega}_i|\varphi_{01}^{(i)}\rangle \tag{B.10}$$

where $\hat{\omega}_i = \hat{1} - N^{-1}|\varphi_0\rangle\langle\varphi_0|\hat{\rho}^{(i)}$, $|\varphi_{01}\rangle = \hat{R}_0\hat{V}|\varphi_0\rangle$, $|\varphi_{01}^{(i)}\rangle = \hat{R}_0\hat{\rho}^{(i)}(\hat{V} - E_{01}^{(i)})|\varphi_0\rangle$. From eq. (V.1.1), $\hat{\rho}^{(i)} = (2)^{-1}[\hat{1} + \varepsilon_i\hat{I}]$; moreover, $|\varphi_{01}^{(i)}\rangle$ can be written as $|\varphi_{01}^{(i)}\rangle = (2)^{-1}[|\varphi_{01}\rangle + \varepsilon_i|\theta\rangle + \varepsilon_i E_{01}^{(i)}|\omega\rangle]$ [see sec. (V.1), particularly what follows eq. (V.1.5)]. Since we are interested in the long-range behaviour of $E_{01}^{(i)}$ (EL-HAV), we shall discard any contribution which drops exponentially to zero at large distances: in particular, the fact that $E_{01}^{(i)}$ is characterized by such behaviour [see eq. (V.1.2')], makes negligible the component to $|\varphi_{01}^{(i)}\rangle$ associated with $|\omega\rangle$ and simplifies the equation for $|\theta\rangle$ [see eqs. (V.1.6)].

The second term on the right-hand side of eq. (B.10) can be rewritten as

$$\Delta^{(i)} = \frac{\varepsilon_i}{2}\langle\varphi_{01}^{as}|\hat{\omega}_i[\hat{I},\hat{V}]\hat{\omega}_i\{|\varphi_{01}^{as}\rangle + \varepsilon_i|\theta^{as}\rangle\}$$

(note that $N \to \frac{1}{2}$); taking into account that $[\hat{I},\hat{V}] = (r_b^{-1} - r_a^{-1})\hat{I}$, the preceding expression for $\Delta^{(i)}$ transforms into

$$\Delta^{(i)} = \frac{\varepsilon_i}{2}\langle\varphi_{01}^{as}|\hat{\omega}_i^{\dagger}(r_b^{-1} - r_a^{-1})\hat{I}\hat{\omega}_i|\varphi_{01}^{as}\rangle + \frac{1}{2}\langle\varphi_{01}^{as}|\hat{\omega}_i^{\dagger}(r_b^{-1} - r_a^{-1})\hat{I}\hat{\omega}_i|\theta^{as}\rangle \tag{B.11}$$

A careful inspection of the various contributions to $\Delta^{(i)}$, so as to identify what "survives" in the long-range region, allows to ascertain that the first term on the right-hand side of eq. (B.11) does not contribute and the second-one contributes only partially, so that, independently of the symmetry,

$$\Delta^{(i)} = \Delta = -\frac{1}{2}\langle\varphi_{01}^{as}|r_a^{-1}\hat{I}|\theta^{as}\rangle = -\frac{1}{2}\langle\hat{I}\,\varphi_{01}^{as}|r_b^{-1}|\theta^{as}\rangle \qquad (B.12)$$

The lowest-order contribution to $|\theta^{as}\rangle$ in its expansion in Legendre polynomials $P_\ell(\cos\theta_b)$ comes from $\ell = 1$; its formal expression [79]

$$\langle r_b,\theta_b,\varphi_b|\theta^{as}\rangle = R^{-2}[(r_b+r_b^2)/3]P_1(\cos\theta_b)\ 1s_b(r_b) + O(R^{-3})$$

along with the preceding results for φ_{01}^{as}, allows to evaluate in a straightforward way the matrix element in eq. (B.12):

$$\Delta = -\frac{13}{24}R^{-4} + O(R^{-6}) \qquad (B.13)$$

From eqs. (B.9), (B.10) and (B.13), after recognizing that $E_{02}^{as}(\text{MS-MA}) = E_{02}^{as}$ (see sec. V.2), we finally obtain

$$E_{02}^{as}(\text{EL-HAV}) = -\frac{41}{24}R^{-4} + O(R^{-6}) \qquad (B.14)$$

APPENDIX C

We propose to examine a number of interesting analytic properties of the dynamic dipole polarizability tensor $\alpha_{ij}(\omega)$. In order to keep the discussion as simple as possible, we confine our attention to the case of a S-state atom, so that only the scalar part $\alpha(\omega)$ of $\alpha_{ij}(\omega)$ is relevant.

From eq. (VII.2.10), one obtains (see Chapters VI, VII)

$$\alpha(\omega) = \frac{2}{3} \sum_k{}' \frac{E_{ko} |\langle \varphi_o | \hat{\underline{\mu}} | \varphi_k \rangle|^2}{E_{ko}^2 - \omega^2} \tag{C.1}$$

where one should emphasize that the spectrum of the atom consists of a discrete portion along with a continuum. Let us now introduce the <u>oscillator strength distribution</u> $df(\varepsilon)/d\varepsilon$ [228,229]

$$\frac{df(\varepsilon)}{d\varepsilon} = \sum_k{}' f_{ko} \delta(E_{ko} - \varepsilon) + \frac{dg(\varepsilon)}{d\varepsilon} \tag{C.2}$$

where $\varepsilon = E - E_o$, $f_{ko} = \frac{2}{3} E_{ko} |\langle \varphi_o | \hat{\underline{\mu}} | \varphi_k \rangle|^2$ is the oscillator strength for the transition $k \leftarrow o$ and

$$\frac{dg(\varepsilon)}{d\varepsilon} = \frac{2}{3} \varepsilon |\langle \varphi_o | \hat{\underline{\mu}} | \varphi(\varepsilon) \rangle|^2$$

represents the oscillator strength for transitions in the continuum in a unit interval at the energy $E = \varepsilon + E_o$; $\sum_k{}' f_{ko} \delta(E_{ko} - \varepsilon)$ corresponds to the discrete portion of the spectrum, as clearly put in evidence by the presence of the Dirac delta functions $\delta(E_{ko} - \varepsilon)$. The dipole polarizability can therefore be rewritten in the form,

$$\alpha(\omega) = \int_0^\infty d\varepsilon \, \frac{(df/d\varepsilon)}{\varepsilon^2 - \omega^2} \tag{C.4}$$

which is a useful starting point for investigating its properties. (We note incidentally that one must take the principal value of the integral as ω falls within an absorption band).

In the normal dispersion region (i.e. for frequencies ω lower than the first resonance E_{10}), the polarizability can be represented by the well known <u>Cauchy dispersion expansion</u>

$$\alpha(\omega) = \sum_{n=0}^{\infty} \alpha_n (\omega^2)^n \tag{C.5}$$

If we perform a corresponding expansion in eq. (C.4), the following integral expression for the Cauchy moments α_n is easily obtained

$$\alpha_n = \int_0^{\infty} d\varepsilon \, \frac{df}{d\varepsilon} \, \varepsilon^{-(2n+2)} \equiv S(-2n-2) \tag{C.6}$$

the notation S(-2n-2) having been introduced in accordance with a rather common practice. Eq. (C.6) can usefully be transformed by changing variable from ε to $u = 1/\varepsilon^2$ [229]:

$$\alpha_n = \int_0^{\infty} u^n \, d\,\phi(u) \tag{C.7}$$

where

$$\phi(u) = \frac{1}{2} \int_0^{u} du \, u^{-\frac{1}{2}} \frac{df}{d\varepsilon} \tag{C.7'}$$

Since the oscillator strength distribution $\frac{df}{d\varepsilon}$ is non-negative definite, from eq. (C.7') it follows that $\phi(u)$ is a non-decreasing function of u and from eq. (C.7) that the Cauchy moments α_n are moments of a non-decreasing distribution. This behaviour characterizes the polarizability series as a <u>series of Stieltjes</u> in the variable $z = -\omega^2$ [229,230]

$$\alpha(z) = \sum_{n=0}^{\infty} \alpha_n (-z)^n \tag{C.8}$$

Among other things, such a series is endowed with the property that the moments α_n must satisfy the determinantal inequalities [229,230]

$$D(n,m) = \begin{vmatrix} \alpha_n & \alpha_{n+1} & \cdots & \alpha_{n+m} \\ \alpha_{n+1} & \alpha_{n+2} & \cdots & \alpha_{n+m+1} \\ \vdots & \vdots & & \\ \alpha_{n+m} & \alpha_{n+m+1} & & \alpha_{n+2m} \end{vmatrix} > 0 \quad (n,m=0,1,\ldots) \qquad (C.9)$$

From eq. (C.7) we can easily establish the radius of convergence of the Cauchy expansion; as known [58], the radius of convergence of a power series is given by $\lim_{n\to\infty} |\alpha_n/\alpha_{n+1}|$, so that from $\int_0^\infty u^{n+1} d\phi(u) \geq E_{10}^{-2} \int_0^\infty u^n \cdot d\phi(u)$ we get $\lim_{n\to\infty} |\alpha_n/\alpha_{n+1}| \leq E_{10}^2$, E_{10} being the smallest value of the variable $\varepsilon = E - E_0$ (in correspondence with the first resonance energy of the system).

The moment $\alpha_0 \equiv S(-2)$ corresponds to the usual static dipole polarizability. α_0 and a limited number of the successive moments $\alpha_1 \equiv S(-4)$, $\alpha_2 \equiv S(-6)$, etc. are susceptible of semiempirical determination, for instance by least-squares fitting a truncated Cauchy expansion to available refractivity or Verdet coefficient data [229]. As an alternative, one can also resort to theoretical calculations, even though the well known difficulties already encountered with α_0 are to be expected.

In addition to the Cauchy moments $S(-2n-2)$, $n = 0,1,\ldots,$ defined by eq. (C.6), which individuate $\alpha(\omega)$ within the radius of convergence of the Cauchy expansion, other moments $S(p)$ are definible; of particular interest are the moments corresponding to $p = -1,0,1,2$, because they can be expressed as expectation values of operators over the ground state wavefunction of the system under consideration, revealing in this way the existence of <u>sum rules</u>. In analogy to eq. (C.6) they are defined as

$$S(p) = \int_0^\infty d\varepsilon \, \frac{df}{d\varepsilon} \, \varepsilon^p \qquad (C.10)$$

After making use of eqs. (C.2), (C.3), it is easily verified for $S(p)$ the expression

$$S(p) = \frac{2}{3} \sum_k E_{ko}^{p+1} \; | \langle \varphi_o \, | \, \hat{\underline{\mu}} \, | \varphi_k \rangle |^2 \qquad (C.11)$$

which is a good starting point for demonstrating its reduction to expectation values of suitable operators over the ground state wavefunction [229]. For $p = -1$, the completeness condition $\sum_k |\varphi_k\rangle\langle\varphi_k| = \hat{1}$ leads immediately to

$$S(-1) = \frac{2}{3} \langle\varphi_0 | \hat{\underline{\mu}}\cdot\hat{\underline{\mu}} | \varphi_0\rangle \tag{C.12}$$

The case $p = 0$, slightly more complicate, is amenable to the following form

$$S(0) = \frac{1}{3} \langle\varphi_0 | [\hat{\mu}\cdot, [\hat{H}_0, \hat{\underline{\mu}}]] | \varphi_0\rangle$$

where $[\hat{A},\hat{B}]$ denotes the commutator of the two operators $\hat{A},\hat{B}$. From $\hat{\underline{\mu}} = \int d\underline{r}\, \underline{r}\, \hat{\rho}(\underline{r}) = -\sum_{j=1}^{N} \hat{\underline{r}}_j$ (N being the number of atomic electrons), it is a standard exercise to show that $[\hat{\underline{\mu}}\cdot, [\hat{H}_0, \hat{\underline{\mu}}]] = 3N$, so that we recover the well known Thomas-Reiche-Kuhn sum rule

$$S(0) = N \tag{C.13}$$

For $p = 1,2$ we find in a similar way [18,229]

$$S(1) = \frac{2}{3} \langle\varphi_0 | (\sum_{j=1}^{N} \hat{\underline{p}}_j)^2 | \varphi_0\rangle \tag{C.14}$$

$$S(2) = \frac{4}{3} \pi N \langle\varphi_0 | \sum_{j=1}^{N} \delta(\underline{r}_j) | \varphi_0\rangle \tag{C.15}$$

where $\underline{p}_j$ is the momentum operator of the i-th electron and $\delta(\underline{r}_j)$ the Dirac delta function. It is not difficult to show that $S(p)$ becomes infinitely large for $p = 3$, so that there are moments of the oscillator strength distribution corresponding to integers p only for $p \le 2$. Actually, for ground state atoms $S(p)$ is infinite for $p \ge 2.5$ [18].

APPENDIX D

In order to derive the basic equations of the time-dependent coupled HF approximation, let us imagine to have a N electron system subjected to a weak external time-dependent perturbation $\hat{V}(t)$ of the form

$$\hat{V}(t) = \hat{v}\, e^{-i\omega t} + \hat{v}^{\dagger}\, e^{i\omega t} \tag{D.1}$$

$\hat{v}$ being an arbitrary one-electron operator quantity.

An approximate state vector $|\phi(t)\rangle$ for the perturbed system in the form of Slater determinant can be built up from a (complete) basis of "unperturbed" spin-orbitals $|\chi_j^o\rangle$, solutions to the HF equations for the ground state of the system in the absence of external field. From eq. (V.3.2) we have in fact

$$|\phi(t)\rangle = \frac{N(t)}{(N!)^{\frac{1}{2}}} \det\, [|\chi_1(1;t)\rangle|\chi_2(2;t)\rangle\ldots|\chi_N(N;t)\rangle] \tag{D.2}$$

where $N(t)$ is a normalization coefficient and

$$|\chi_\alpha(t)\rangle = |\chi_\alpha^o\rangle + \sum_r |\chi_r^o\rangle\, C_{r\alpha}(t), (1 \leq \alpha \leq N) \tag{D.3}$$

(we recall that r labels virtual, i.e. unoccupied, spinorbitals in the unperturbed HF ground state determinant). If we substitute eq. (D.3) into eq. (D.2), after retaining terms which are at most quadration in the coefficients $C_{r\alpha}(t)$ we get

$$|\phi(t)\rangle = N(t)\{|\phi_o\rangle + \sum_{r\alpha} C_{r\alpha}(t)\,|\phi_\alpha^r\rangle + \frac{1}{2}\sum_{r\alpha}\sum_{s\beta} C_{r\alpha}(t)\, C_{s\beta}(t)\,|\phi_{\alpha\beta}^{rs}\rangle\} \tag{D.4}$$

where $|\phi_o\rangle \equiv (N!)^{-\frac{1}{2}} \det\, [|\chi_1^o(1)\rangle|\chi_2^o(2)\rangle\ldots|\chi_N^o(N)\rangle]$ is the (normalized) Slater determinant approximate wavefunction for the ground state of the system in the absence of external perturbations, while $|\phi_\alpha^r\rangle$, $|\phi_{\alpha\beta}^{rs}\rangle$,.... are (normalized) singly-, doubly-,.... excited configurations (see also sec. V.3).

The unknown quantities $N(t)$ and $C_{r\alpha}(t)$ of the approximate solution (D.4) to the time-dependent Schrödinger equation can be determined from the so-called Frenkel's variational principle [237]

$$\langle\delta\phi(t)|\hat{H}_0 + \hat{V}(t) - i\frac{\partial}{\partial t}|\phi(t)\rangle = 0 \tag{D.5}$$

if we vary $|\phi(t)$ with respect to $N(t)$ and $C_{r\alpha}(t)$. If we put, for simplicity, $|\phi(t)\rangle = N(t)|\psi(t)\rangle$, so that $|\delta\phi(t)\rangle = \delta N(t)|\psi(t)\rangle + N(t)|\delta\psi(t)\rangle$, from eq. (D.5) we obtain the following resolving conditions

$$\frac{\dot{N}(t)}{N(t)} = -i\frac{\langle\psi(t)|\hat{H}_0 + \hat{V} - i\frac{\partial}{\partial t}|\psi(t)}{\langle\psi(t)|\psi(t)\rangle} \tag{D.6}$$

$$N(t)\langle\delta\psi(t)|\hat{H}_0 + \hat{V}(t) - i\frac{\partial}{\partial t}|\psi(t)\rangle = i\dot{N}(t)\langle\delta\psi(t)|\psi(t)\rangle \tag{D.6'}$$

Eq. (D.6) can be integrated without any difficulties

$$N(t) = N(t_0)\exp\{-i\int_{t_0}^{t} dt' \frac{\langle\psi(t')|\hat{H}_0+\hat{V}(t')-i\frac{\partial}{\partial t'}|\psi(t')\rangle}{\langle\psi(t')|\psi(t')\rangle}\}$$

$N(t_0)$ being the value of $N(t)$ at an arbitrary time t_0, and one recognized easily that $\langle\phi(t)|\phi(t)\rangle = |N(t)|^2\langle\psi(t)|\psi(t)\rangle = |N(t_0)|^2\langle\psi(t_0)|\psi(t_0)\rangle = \langle\phi(t_0)|\phi(t_0)\rangle$, i.e. $|\phi(t)\rangle$ conserves its norm at any time. From eq. (D.6), eq. (D.6') is immediately cast into the form (not involving (t)),

$$\langle\delta\psi(t)|\hat{H}_0+\hat{V}(t)-i\frac{\partial}{\partial t}|\psi(t)\rangle - \frac{\langle\psi(t)|\hat{H}_0+\hat{V}-i\frac{\partial}{\partial t}|\psi(t)\rangle}{\langle\psi(t)|\psi(t)\rangle}\langle\delta\psi(t)|\psi(t)\rangle = 0 \tag{D.7}$$

which is the starting point for determining the coefficients in the trial wavefunction (D.4), with $|\psi(t)\rangle \equiv |\phi_0\rangle + \sum_{r\alpha} C_{r\alpha}(t)|\phi_\alpha^r\rangle + \frac{1}{2}\sum_{r\alpha}\sum_{s\beta} C_{r\alpha}(t)\, C_{s\beta}(t)|\phi_{\alpha\beta}^{rs}\rangle$. If we keep in eq. (D.7) only terms which are of the order $\delta C^* C$, it is a simple task to derive the following result

$$\langle\phi_\alpha^r|\hat{V}(t)|\phi_o\rangle + \sum_{\beta s}\left[\langle\phi_\alpha^r|\hat{H}_o|\phi_\beta^s\rangle C_{s\beta}(t) + C^*_{s\beta}(t)\langle\phi_{\alpha\beta}^{rs}|\hat{H}_o|\phi_o\rangle\right]$$
$$-i\,\dot{C}_{r\alpha}(t) - E_{HF}\,C_{r\alpha}(t) = 0 \qquad (D.8)$$

where $\dot{C}_{r\alpha} = dC_{r\alpha}/dt$, $E_{HF} = \langle\phi_o|\hat{H}_o|\phi_o\rangle$, the HF energy of the unperturbed ground state of the system and use has been made of the Brillouin's theorem, $\langle\phi_o|\hat{H}_o|\phi_\alpha^r\rangle = 0$ [7,33]. The set of eqs. (D.8) for the time-dependent coefficients $C_{r\alpha}(t)$ can be further transformed if we consider the form of the perturbation term $\hat{V}(t)$, eq. (D.1); if we put

$$C_{r\alpha}(t) = Y_{r\alpha}(\omega)e^{-i\omega t} + Z^*_{r\alpha}(\omega)e^{i\omega t} \qquad (D.9)$$

the following set of equations for the unknown coefficients $Y(\omega)$, $Z(\omega)$ results

$$\sum_{\beta s}\left[\langle\phi_\alpha^r|\hat{H}_o - E_{HF}|\phi_\beta^s\rangle Y_{s\beta} + \langle\phi_{\alpha\beta}^{rs}|\hat{H}_o|\phi_o\rangle Z_{s\beta}\right] - \omega Y_{r\alpha}(\omega) = -\langle\phi_\alpha^r|\hat{v}|\phi_o\rangle$$
$$\sum_{\beta s}\left[\langle\phi_\alpha^r|\hat{H}_o - E_{HF}|\phi_\beta^s\rangle Z_{s\beta} + \langle\phi_{\alpha\beta}^{rs}|\hat{H}_o|\phi_o\rangle Y_{s\beta}\right] + \omega Z_{r\alpha}(\omega) = -\langle\phi_\alpha^r|\hat{v}|\phi_o\rangle \qquad (D.10)$$

These equations can be expressed in a very compact form on introducing the self-adjoint supermatrices $\mathbb{A}$ and $\mathbb{B}$, respectively of elements

$$\mathbb{A}_{r\alpha,s\beta} = \langle\phi_\alpha^r|\hat{H}_o - E_{HF}|\phi_\beta^s\rangle, \quad \mathbb{B}_{r\alpha,s\beta} = \langle\phi_{\alpha\beta}^{rs}|\hat{H}_o|\phi_o\rangle \qquad (D.11)$$

and the (column) supervector $|\mathbb{Y}\rangle$, $|\mathbb{Z}\rangle$ of elements $Y_{r\alpha}(\omega)$ and $Z_{r\alpha}(\omega)$; eqs. (D.10) can be written down as

$$(\mathbb{A} - \omega\mathbb{1})|\mathbb{Y}\rangle + \mathbb{B}|\mathbb{Z}\rangle = -|\boldsymbol{v}\rangle$$
$$(\mathbb{A} + \omega\mathbb{1})|\mathbb{Z}\rangle + \mathbb{B}|\mathbb{Y}\rangle = -|\boldsymbol{v}\rangle \qquad (D.12)$$

$\mathbb{1}$ being the unit supermatrix of proper dimensionality and $|\boldsymbol{v}\rangle$ the (column) supervector of elements $\langle\phi_\alpha^r|\hat{v}|\phi_o\rangle$. Eqs. (D.12) are the basic equations to be solved in the coupled H.F. theory.

The matrix $\mathbb{A}$ is recognized as the matrix to be diagonalized in a calculation of excited states which involves only singly-excited configurations in addition to the H.F. ground state, while $\mathbb{B}$ introduces doubly-excited configurations, which take into account lowest order correlation effects. From eq. (D.4) one should also note that doubly-excited configurations enter the coupled H.F. theory in a peculiar way, their coefficients $C_{r\alpha,s\beta}$ being restricted to the form $C_{r\alpha}\,C_{s\beta}$.

To the end of better appreciating the theory, we shall now consider the realistic case of a N electron system subjected to a monochromatic, homogeneous external electric field $\underline{F}$ polarized along Oz, so that in eq. (D.1) $\hat{v}=\hat{v}^\dagger=-F_z\,\hat{\mu}_z$, $\hat{\mu}_z$ being the ordinary electric dipole moment operator. The expectation value $\langle\hat{\mu}_z\rangle$ for the dipole moment of the system is simply given by

$$\langle\hat{\mu}_z\rangle_t = \langle\hat{\mu}_z\rangle_o + \left[\langle\boldsymbol{\mu}_z|\mathbb{Y}+\mathbb{Z}\rangle e^{-i\omega t} + \text{c.c.}\right] \qquad \text{(D.13)}$$

where $\langle\hat{\mu}_z\rangle_o = \langle\phi_o|\hat{\mu}_z|\phi_o\rangle$ is the (approximate) H.F. value of the permanent dipole moment, and the quantity in square brackets is the induced electric dipole moment, expressed in terms of the supervectors $|\mathbb{Y}+\mathbb{Z}\rangle$ and $|\boldsymbol{\mu}_z\rangle$ ($|\boldsymbol{\mu}_z\rangle$ has real elements $\langle\phi_o|\hat{\mu}_z|\phi_\alpha^r\rangle$). Since only the quantity $|\mathbb{Y}+\mathbb{Z}\rangle$ is requested in eq. (D.13), we can manipulate eqs. (D.12) to give

$$(\mathbb{A}+\mathbb{B})|\mathbb{Y}+\mathbb{Z}\rangle - \omega|\mathbb{Y}-\mathbb{Z}\rangle = 2\,F_z|\boldsymbol{\mu}_z\rangle$$

$$(\mathbb{A}-\mathbb{B})|\mathbb{Y}-\mathbb{Z}\rangle - \omega|\mathbb{Y}+\mathbb{Z}\rangle = \mathbb{0}$$

and therefore

$$\left[(\mathbb{A}-\mathbb{B})(\mathbb{A}+\mathbb{B}) - \omega^2\mathbb{1}\right]|\mathbb{Y}+\mathbb{Z}\rangle = +2\,F_z(\mathbb{A}-\mathbb{B})|\boldsymbol{\mu}_z\rangle \qquad \text{(D.15)}$$

Thus we obtain formally

$$|\mathbb{Y} + \mathbb{Z}\rangle = 2F_z\left[(\mathbb{A}-\mathbb{B})(\mathbb{A}+\mathbb{B}) - \omega^2\mathbb{1}\right]^{-1}(\mathbb{A}-\mathbb{B})\,|\boldsymbol{\mu}_z\rangle$$

$$= 2F_z\left[(\mathbb{A}+\mathbb{B}) - \omega^2(\mathbb{A}-\mathbb{B})^{-1}\right]^{-1}|\boldsymbol{\mu}_z\rangle$$

and consequently $\langle\hat{\mu}_z\rangle_t$, eq. (D.13), becomes

$$\langle\hat{\mu}_z\rangle_t = \langle\hat{\mu}_z\rangle_0 + 2F_z(e^{-i\omega t} + e^{i\omega t})\langle\boldsymbol{\mu}_z|\left[(\mathbb{A}-\mathbb{B})(\mathbb{A}+\mathbb{B}) - \omega^2\mathbb{1}\right]^{-1} \cdot (\mathbb{A}-\mathbb{B})\,|\boldsymbol{\mu}_z\rangle \tag{D.16}$$

The zz-component of the dynamic dipole polarizability tensor $\alpha_{zz}(\omega)$ of the system in the coupled H.F. theory is therefore

$$\alpha_{zz}(\omega) = 2\langle\boldsymbol{\mu}_z|\left[(\mathbb{A}-\mathbb{B})(\mathbb{A}+\mathbb{B}) - \omega^2\mathbb{1}\right]^{-1}(\mathbb{A}-\mathbb{B})\,|\boldsymbol{\mu}_z\rangle \tag{D.17}$$

A further interesting development of this result is possible in terms of the eigenvalues and eigenvectors to the not self-adjoint matrix $(\mathbb{A}-\mathbb{B})(\mathbb{A}+\mathbb{B})$:

$$(\mathbb{A}-\mathbb{B})(\mathbb{A}+\mathbb{B})\,|\mathbb{C}_n\rangle = \omega_n^2\,|\mathbb{C}_n\rangle \tag{D.18}$$

If we consider this eigenvalue problem in conjunction with the associated one

$$(\mathbb{A}+\mathbb{B})(\mathbb{A}-\mathbb{B})\,|\mathbb{D}_n\rangle = \Omega_n^2\,|\mathbb{D}_n\rangle \tag{D.18'}$$

it can easily be demonstrated that $\Omega_n^2 = \omega_n^2$ and $\langle\mathbb{D}_n|\mathbb{C}_m\rangle = \langle\mathbb{C}_m|\mathbb{D}_n\rangle = \delta_{nm}$ ($\{|\mathbb{D}_n\rangle\}$ and $\{|\mathbb{C}_n\rangle\}$ are said to constitute a biorthonormal set). The completeness relation for a biorthonormal set is then written as

$$\sum_m |\mathbb{C}_n\rangle\langle\mathbb{D}_n| = \sum_n |\mathbb{D}_n\rangle\langle\mathbb{C}_n| = \mathbb{1} \tag{D.19}$$

which provides the following spectral representation for the operator

$(\mathbb{A}-\mathbb{B})(\mathbb{A}+\mathbb{B})$

$$(\mathbb{A}-\mathbb{B})(\mathbb{A}+\mathbb{B}) = \sum_n \omega_n^2 \,|\mathbb{C}_n\rangle\langle\mathbb{D}_n| \tag{D.20}$$

The polarizability $\alpha(\omega)$, eq. (D.17), can therefore be expressed in the form

$$\alpha_{zz}(\omega) = 2\sum_n \frac{\langle\mathbb{\mu}_z|\mathbb{C}_n\rangle\langle\mathbb{D}_n|(\mathbb{A}-\mathbb{B})|\mathbb{\mu}_z\rangle}{\omega_n^2-\omega^2} \tag{D.21}$$

which suggests the interpretation that the eigenvalues ω_n^2 to $(\mathbb{A}-\mathbb{B})(\mathbb{A}+\mathbb{B})$ are to be regarded as (approximate) square excitation energies of the (unperturbed) system. Without developing further this subject, we limit ourselves to state that solving the eigenvalue problem expressed by eqs. (D.18), (D.18') is one of the ways suggested for tackling the coupled H.F. problem.

REFERENCES

[1] Hirschfelder, J.O., Curtiss, C.F. and Bird, R.B., Molecular Theory of Gases and Liquids, Wiley, N.Y., 1964.

[2] Hirschfelder, J.O. (ed.), Intermolecular Forces, Adv. Chem.Phys., XII, Interscience, 1967.

[3] Buckingham, A.D. and Utting, B.D. in Ann. Rev. Phys. Chem., 21, 287 (1970).

[4] Margenau, H. and Kestner, N.R., Theory of Intermolecular Forces, 2-nd ed., Pergamon Press, London, 1971.

[5] Certain, P.R. and Bruch, L.W. in Intern. Rev. Science, Phys.Chem. Series 1, Vol. 1, Butterworths, London, 1972.

[6] Pullman, B. (ed.), Intermolecular Interactions: from Diatomics to Biopolymers, Wiley, N.Y., 1978.

[7] Pilar, F.K., Elementary Quantum Chemistry, McGraw-Hill, N.Y., 1968.

[8] Löwdin, P.O., Rev. Mod. Phys., 34, 520 (1962).

[9] Löwdin, P.O., Rev. Mod. Phys., 39, 259 (1967).

[10] Löwdin, P.O., Intern. J. Quantum Chem., 2, 867 (1968).

[11] Johnston, D.F., Rep. Progr. Phys., 23, 66 (1960).

[12] Hamermesh, M., Group Theory, Addison-Wesley, Reading, Mass., 1962.

[13] Kaplan, I.G., Symmetry of Many-Electron System, Academic Press, N.Y., 1975.

[14] Chisholm, C.D.H., Group Theoretical Techniques in Quantum Chemistry, Academic Press, N.Y. 1976.

[15] Schensted, I.V., A Course on the Application of Group Theory to Quantum Mechanics, Neo Press, Peaks Island, Maine, 1976.

[16] Coleman, A.J. in Adv. Quantum Chem., 4, 83 (1968).

[17] Claverie, P., Intern. J. Quantum Chem., 5, 273 (1971).

[18] Hirschfelder, J.O., Byers Brown, W. and Epstein, S.T. in Adv. Quantum Chem., 1, 255 (1964).

[19] Wilcox, C.H. (ed.), Perturbation Theory and its Applications to Quantum Mechanics, J.Wiley, N.Y., 1966.

[20] Löwdin, P.O., J. Mol. Spectry 10, 12 (1963).

[21] Löwdin, P.O., J.Mol. Spectry 13, 326 (1964).

[22] Löwdin, P.O., J. Math. Phys., 3, 969 (1962).

[23] Löwdin, P.O., J. Math. Phys. 3, 1171 (1962)

[24] Löwdin, P.O., J. Mol. Spectry 14, 112 (1964).

[25] Löwdin, P.O., J. Mol. Spectry 14, 119 (1964).

[26] Löwdin, P.O., J. Mol. Spectry 14, 131 (1964).

[27] Löwdin, P.O., J. Math. Phys. 6, 1341 (1965).

[28] Löwdin, P.O., Phys. Rev. 139, A357 (1965).

[29] Löwdin, P.O., in Perturbation Theory and its Applications to Quantum Mechanics, (Wilcox, C.H., ed.), J. Wiley, N.Y., 1966.

[30] Löwdin, P.O., J. Chem. Phys. 43, 5175 (1965).

[31] Löwdin, P.O., Intern. J. Quantum Chem. 2 S, 137 (1968).

[32] Löwdin, P.O. and Goscinski, O., Intern. J. Quantum Chem., 5, 685 (1971).

[33] Hurley, A.C., Introduction to the Electron Theory of Small Molecules, Academic Press, N.Y., 1976.

[34] van der Avoird, A., Chem. Phys. Letters, 1, 24 (1967).

[35] Alonso, M. and Valk, H., Quantum Mechanics: Principles and Applications, Addison-Wesley, Reading, Mass., 1973.

[36] Murrell, J.N. and Shaw, G., J. Chem. Phys. 46, 1768 (1967).

[37] Chipman, D.M., Bowman, J.D. and Hirschfelder, J.O., J. Chem.Phys. 59, 2830 (1973).

[38] Amos, A.T. and Musher, J.I., Chem. Phys. Letters 1, 149 (1967).

[39] Musher, J.I. and Amos, A.T., Phys. Rev. 164, 31 (1967).

[40] Brändas, E. and Goscinski, O., J. Chem. Phys. 51, 975 (1969).

[41] Snook, I.K. and Spurling, T.H., J. Chem. Soc. Faraday Trans. II, 71, 852 (1975).

[42] van der Avoird, A., J. Chem. Phys. 47, 3649 (1967).

[43] Hirschfelder, J.O., Chem. Phys. Letters 1, 363 (1967).

[44] van der Avoird, A., Chem. Phys. Letters 1, 411 (1967).

[45] Johnson, R.E. and Epstein, S.T., Chem. Phys. Letters 1, 599 (1968).

[46] Eisenschitz, R. and London, F., Z. Phys. 60, 491 (1930).

[47] Kemble, E.C., Fundamental Principles of Quantum Mechanics, Dover Publ., Inc., N.Y., (1958).

[48] Kirtman, B., J. Chem. Phys. 49, 3890 (1968).

[49] Micha, D.A., J. Chem. Phys. 48, 3639 (1968).

[50] Lindner, P. and Löwdin, P.O., Intern. J. Quantum. Chem. 2S, 161 (1968).

[51] Goscinski, O., Intern. J. Quantum Chem. 2, 761 (1968).

[52] Löwdin, P.O., Intern. J. Quantum Chem. 4, 231 (1971).

[53] Abdulnur, S. and Öhrn, Y., Chem. Phys. Letters 6, 502 (1970).

[54] Abdulnur, S., Intern. J. Quantum Chem. 5, 525 (1971).

[55] Hirschfelder, J.O., Chem. Phys. Letters 1, 325 (1967).

[56] Musher, J.I., Rev. Mod. Phys. 39, 203 (1967).

[57] Ahlrichs, R., Chem. Phys. Letters 18, 67 (1973).

[58] Whittaker, E.T. and Watson, G.N., A Course of Modern Analysis, Cambridge University Press (1965).

[59] Robinson, P.D., Proc. Phys. Soc. 78, 537 (1961).

[60] Claverie, P., Intern. J. Quantum Chem. 3, 349 (1969).

[61] Certain, P.R. and Byers Brown, W., Intern. J. Quantum Chem. 6, 131 (1972).

[62] Ahlrichs, R. and Claverie, P., Intern. J. Quantum Chem. 6, 1001 (1972).

[63] Gasiorowicz, S., Quantum Physics, J. Wiley, N.Y. (1974).

[64] Chalasinski, G., Jeziorski, B. and Szalewicz, K., Intern. J. Quantum Chem. 11, 247 (1977).

[65] Peek, J.M., J. Chem. Phys. 43, 3004 (1965).

[66] Kolos, W. and Wolniewicz, L., Chem. Phys. Letters 24, 457 (1974).

[67] Courant, R. and Hilbert, D., Methods of Mathematical Physics, vol. I, Interscience, N.Y. (1953).

[68] Mattis, D.C., The Theory of Magnetism, Harper & Row Publ., N.Y. (1965).

[69] Landau, L. and Lifchitz, E., Mécanique Quantique, Eds Mir, Moscow (1966).

[70] Razi Naqvi, K. and Byers Brown, W., Intern. J. Quantum Chem. 6, 271 (1972).

[71] Herring, C. and Flicker, M., Phys. Rev. 134, A 362 (1964).

[72] Ahlrichs R., Theoret. Chim. Acta 41, 7 (1976).

[73] Dalgarno, A. and Lynn, N., Proc. Phys. Soc. Lond. A69, 821 (1956).

[74] Lyon, W.D., Matcha, R.L., Sanders, W.A., Meath, W.J. and Hirschfelder, J.O., J. Chem. Phys. 43, 1095 (1965).

[75] Mc Quarrie, D.A. and Hirschfelder, J.O., J. Chem. Phys. 47, 1775 (1967).

[76] van der Avoird, A., Chem. Phys. Letters 1, 429 (1967).

[77] Sanders, W.A., J. Chem. Phys. 51, 491 (1969).

[78] Laughlin, C. and Amos, A.T., J. Chem. Phys. 55, 4837 (1971).

[79] Chipman, D.M. and Hirschfelder, J.O., J. Chem. Phys. 59, 2838 (1973).

[80] Chalasinski, G. and Jeziorski, B., Intern. J. Quantum Chem. 7, 63 (1973).

[81] Chalasinski, G. and Jeziorski, B., Intern. J. Quantum Chem. 7, 745 (1973).

[82] Magnasco, V., Battezzati, M. and Figari, G., J. Chem. Soc. Faraday II 72, 22 (1976).

[83] Battezzati, M. and Magnasco, V., J. Chem. Soc. Faraday II 72, 508 (1976).

[84] Eyring, H., Walter, J. and Kimball, G., Quantum Chemistry, Wiley N.Y., 1944.

[85] Levine, I.N., Quantum Chemistry, Allyn and Bacon, Boston, 1970.

[86] Dalgarno, A. and Lynn, N., Proc. Phys. Soc. Lond. 70, 223 (1957).

[87] Robinson, P.D., Proc. Phys. Soc. Lond. 71, 828 (1958).

[88] Certain, P.R. and Hirschfelder, J.O., J. Chem. Phys. 52, 5992 (1970).

[89] Rosenthal, C.M., J. Chem. Phys. 53, 3709 (1970).

[90] Hylleraas, E.A., Z. Physik 65, 209 (1930).

[91] Murrell, J.N., Randic, M. and Williams, D.R., Proc. Roy Soc. (London) A70, 566 (1965).

[92] Kreek, H. and Meath, W.J., J. Chem. Phys. 50, 2289 (1969).

[93] Brooks, F.C., Phys. Rev. 86, 92 (1952).

[94] Roe, G.M., Phys. Rev. 88, 659 (1952).

[95] Dalgarno A. and Lewis, J.T., Proc. Phys. Soc. A69, 57, 59 (1956).

[96] Cusachs, L.C., Phys. Rev. 125, 561 (1962).

[97] Dennery, P. and Krzywicki, A., Mathematics for Physicists, Harper & Row Publ., N.Y. (1967).

[98] Goscinski, O. and Brändas, E., Chem. Phys. Letters 2, 299 (1968).

[99] Brändas, E. and Goscinski, O., Intern. J. Quantum Chem. 35, 383 (1970).

[100] Hirschfelder, J.O. and Certain, P.R., Chem. Phys. Letters 2, 539 (1968).

[101] Feenberg, E., Phys. Rev. 103, 1116 (1956).

[102] Kumar, K., Perturbation Theory and the Nuclear Many-Body Problem, North-Holland Publ. (1962).

[103] van Duijneveldt - van de Rijdt, J.G.C.M. and van Duijneveldt, F.B., Chem. Phys. Letters, 17, 425 (1972).

[104] See Claverie's contribution in ref. [6].

[105] Chalasinski, G., Jeziorski, B., Andzelm, J. and Szalewicz, K., Mol. Phys. 33, 971 (1977).

[106] Schaefer, H.F., The Electronic Structure of Atoms and Molecules, Addison-Wesley, Reading, Mass., 1972.

[107] Musher, J.I. in Intern. Rev. Science, Phys. Chem. Series 1, Vol. 1, Butterworth, London, 1972.

[108] Hurley, A.C., Electron Correlation in Small Molecules, Academic Press, N.Y., 1976.

[109] Freeman, D.L. and Karplus, M., J. Chem. Phys. 64, 2641 (1976).

[110] Pople, J.A., Binkley, J.S. and Seeger, R., Inter. J. Quantum Chem. S10, 1 (1976).

[111] Musher, J.I. and Schulman, J.M., Phys. Rev. 173, 93 (1968).

[112] Kelly, H.P., Phys. Rev. 136, B896 (1964).

[113] Kelly, H.P., Adv. Theoret. Phys. 2, 75 (1968).

[114] Silvestone, H.J. and Yin, M.L., J. Chem. Phys., 49, 2026 (1968).

[115] Huzinaga, S. and Arnau, C., Phys. Rev. A1, 1285 (1970).

[116] Møller, C. and Plesset, M.S., Phys. Rev. 46, 618 (1934).

[117] Bartlett, R.J. and Silver, D.M., Intern. J. Quantum Chem. S9, 183 (1975).

[118] Epstein, P.S., Phys. Rev. 28, 695 (1926).

[119] Nesbet, R.K., Proc. Roy Soc. A230, 312, 922 (1955).

[120] Claverie, P., Diner, S. and Malrieu, J.P., Intern. J. Quantum Chem. 1, 751 (1967).

[121] Broussard, J.T. and Kestner, N.R., J. Chem. Phys. 53, 1507 (1970).

[122] Broussard, J.T. and Kestner, N.R., J. Chem. Phys. 58, 3593 (1973).

[123] Dalgarno, A., Advan. Phys. 11, 281 (1962).

[124] Stevens, R.M., Pitzer, R.M. and Lipscomb, W.N., J. Chem. Phys. 38, 550 (1963).

[125] Langhoff, P.W., Karplus, M. and Hurst, R.P., J. Chem. Phys. 44, 505 (1966).

[126] Caves, T.C. and Karplus, M., J. Chem. Phys. 50, 3649 (1969).

[127] Magnasco, V., Battezzati, M. and Austi, R., Chem. Phys. Letters 51, 375 (1977).

[128] Liu, B. and Mc Lean, A.D., J. Chem. Phys. 59, 4557 (1973).

[129] Ostlund, N.S. and Merrifield, D.L., Chem. Phys. Letters 39, 612 (1976).

[130] Burton, P.G., J. Chem. Phys. 67, 4696 (1977).

[131] Longuet-Higgins, H. C., Proc. Roy Soc. (London) A235, 537 (1956).

[132] Longuet-Higgins, H. C. and Salem, L., Proc. Roy Soc. (London) A259, 433 (1961).

[133] Jackson, J.D., Classical Electrodynamics (2-nd ed.), J. Wiley, N.Y., 1975.

[134] Löwdin, P.O., Phys. Rev. 97, 1474 (1955).

[135] Mc Weeny, R., Rev. Mod. Phys. 32, 335 (1960).

[136] Mc Weeny, R. and Sutcliffe, B.T., Methods of Molecular Quantum Mechanics, Academic Press, N.Y., 1969.

[137] Fukui, K. and Yamabe, T., Intern. J. Quantum. Chem. 2, 359 (1968).

[138] Gordon, R.G. and Kim, Y.S., J.Chem.Phys. 56, 3122 (1972).

[139] Buehler, R.J. and Hirschfelder, J.O., Phys. Rev. 83, 628 (1951).

[140] Buehler, R.J. and Hirschfelder, J.O., Phys. Rev. 85, 149 (1952).

[141] Ng, K.C., Meath, W.J. and Allnatt, A.R., Mol. Phys. 32, 177 (1976).

[142] Rose, M.E., Elementary Theory of Angular Momentum, J.Wiley, 1957.

[143] Brink, D.M. and Satchler, G.R., Angular Momentum, Oxford U.P., 1962

[144] Dalgarno, A. and Davison, W.D., Advances in Atomic and Molecular Physics, 2, 1 (1966).

[145] Wormer, P.E.S., Intermolecular Forces and the Group Theory of Many-Body Systems, Thesis, Nijmegen, 1975.

[146] Rose, M.E., J.Math. and Phys. 37, 215 (1958).

[147] Kreek, H., Pan, Y.H. and Meath, W.J., Mol. Phys. 19, 513 (1970).

[148] Pan Y.H. and Meath, W.J., Mol. Phys. 20, 873 (1971).

[149] Riera, A. and Meath, W.J., Mol. Phys. 24, 1407 (1972).

[150] Bukta, J.F. and Meath, W.J., Intern. J. Quantum. Chem. VI, 1045 (1972).

[151] Clementi, E., J.Chem.Phys. 46, 3851 (1967).

[152] Davidson, E.R. and Jones, L.L., J.Chem. Phys. 37, 2966 (1962).

[153] Jeziorski, B., Bulski, M. and Piela, L., Intern. J. Quantum Chem. X, 281 (1975).

[154] Williams, D.R., Schaad, L.J. and Murrell, J.N., J.Chem.Phys. 47, 4916 (1967).

[155] van Duijneveldt-van der Rijdt, J.G.C.M. and van Duijneveldt, F.B., Chem. Phys. Letters 2, 565 (1968).

[156] Dacre, P.D. and Mc Weeny, R., Proc. Roy. Soc. (London) A 317, 435 (1970).

[157] Conway, A. and Murrell, J.N., Mol. Phys. 23, 1143 (1972).

[158] Riera, A. and Meath, W.J., Intern. J. Quantum. Chem. VII, 959, (1973).

[159] Wormer, P.E.S., Mulder, F. and van der Avoird, A., Intern. J. Quantum Chem. XI, 959 (1977).

[160] Reed, K.A. and Wharton, L., J.Chem.Phys., 66, 3399 (1977).

[161] Van Kranendonk, J., Physica 73, 156 (1974).

[162] Le Roy, R.J. and Van Kranendonk, J., J.Chem.Phys. 61, 4750 (1974).

[163] Dunker, A.M. and Gordon, R.G., J.Chem.Phys. 64, 354 (1976).

[164] Dunker, A.M. and Gordon, R.G., J.Chem.Phys. 68, 700 (1978).

[165] Koide,A., J.Phys. B 11, 633 (1978).

[166] Oka, T., Advances in Atomic and Molecular Physics, 9, 127 (1973).

[167] Landau, L. and Lifchitz, E., Physique Statistique, Eds. Mir, Moscow (1967).

[168] Casimir, H.B.G. and Polder, D., Phys. Rev. 73, 360 (1948).

[169] Power, E.A. and Zienau, S., Nuovo Cimento 6, 7 (1957).

[170] Aub, M.R., Power, E.A. and Zienau, S., Phil. Mag. 2, 571 (1957).

[171] Dzialoshinskii, I.E., Sov. Phys. JETP 3, 977 (1957).

[172] McLachlan, A.D., Proc. Roy. Soc. (London) A271, 387 (1963).

[173] Mavroyannis, C. and Stephen, M.J., Mol. Phys. 5, 629 (1962).

[174] Feinberg, G. and Sucher, J., Phys. Rev. 139, 1619 (1965).

[175] Power, E.A. in ref. [2].

[176] Linder, B. and Rabenold, D.A., Adv. Quantum Chem. 6 203 (1972).

[177] Meath, W.J. and Hirschfelder, J.O., J.Chem.Phys. 44, 2130 (1966).

[178] Johnson, R.E., Epstein, S.T., and Meath, W.J., J.Chem.Phys., 47, 1271 (1967).

[179] Getzin, P.M. and Karplus, M., J.Chem.Phys., 52, 2100 (1970).

[180] Starkschall, G. and Gordon, R.G., J.Chem.Phys. 67, 3213 (1972).

[181] Young, R.H., Intern. J. Quantum Chem. 9, 47 (1975).

[182] McLean, A.D. and Yoshimine, M., J.Chem.Phys. 47, 1927 (1967).

[183] Victor, G.A. and Dalgarno, A., J.Chem.Phys. 53, 1316 (1970).

[184] Langhoff, P.W., Gordon, R.G. and Karplus, M., J.Chem.Phys., 55, 2126 (1971).

[185] Stogryn, D.E. and Stogryn, A.P., Mol. Phys. 11, 371 (1966).

[186] Miller, T.M. and Bederson, B., Advances in Atomic and Molecular Physics 13, 1 (1977).

[187] Dalgarno, A. and Kingston,A.E., Proc. Phys. Soc. (London) 73, 455 (1959).

[188] Cohen, M., Can. J. Phys. 45, 3387 (1967).

[189] Hyman, H.A., J.Chem. Phys. 61, 4063 (1974).

[190] Buckingham, A.D. and Orr, B.J., Quart. Rev. 21, 195 (1967).

[191] Cohen, H.D. and Roothaan, C.C.J., J.Chem.Phys. 43, S34 (1965).

[192] Chang, E.S., Pu, R.T., and Das, T.P., Phys. Rev. 174, 16 (1968).

[193] Werner, H.J. and Meyer, W., Phys. Rev. A13, 13 (1976).

[194] Werner, H.J. and Meyer, W., Mol. Phys. 31, 855 (1976).

[195] Dalgarno, A. and McIntyre, H.A.J., Proc.Roy. Soc. (London) 85, 47 (1965).

[196] Arrighini, G.P. and Guidotti, C., Mol.Phys. 28, 273 (1974).

[197] Mukherji, P.K., Moitra, R.K. and Muckherji, A., Int.J.Quantum Chem. 5, 637 (1971).

[198] Gutschick, V.P. and McKoy, V., J.Chem.Phys. 58, 2397 (1973).

[199] Teachout, R.R. and Pack, R.T., At. Data 3, 195 (1971).

[200] Arrighini, G.P., Biondi, F. and Guidotti, C., Phys. Rev. A8, 577 (1973).

[201] Sitter, R.E. and Hurst, R.P., Phys.Rev. A5, 5 (1972).

[202] Tuan, D.F.T. and Davidz, A., J.Chem.Phys. 55, 1286 (1971).

[203] Arrighini, G.P., Guidotti, C. and Salvetti, O., J.Chem.Phys. 52, 1037 (1970).

[204] Moccia, R., Theor.Chim.Acta 8, 192 (1967).

[205] Arrighini,G.P., Guidotti, C., Maestro, M., Moccia, R. and Salvetti, O, J.Chem.Phys. 49, 2224 (1968).

[206] Langhoff, P.W. and Hurst, R.P., Phys.Rev. 139A, 1415 (1965).

[207] Harrison, J.F., J.Chem.Phys. 49, 3321 (1968).

[208] Liebman, S.P. and Moskowitz, J.W., J.Chem.Phys. 54, 3622 (1971).

[209] Stevens, W.J. and Billingsley, F.P., Phys.Rev. A8, 2236 (1973).

[210] Wahl, A.C. and Das, G., Adv. Quantum Chem. 5, 261 (1970).

[211] Das, G. and Wahl, A.G., J.Chem.Phys. 56, 1769 (1972).

[212] Magnasco, V. and Amelio, M., J.Chem.Phys. 69, 4706 (1978).

[213] Meyer, W., J.Chem.Phys. 58, 1017 (1973).

[214] Lahiri, J. and Mukherji, A., Phys.Rev. 141, 428 (1966).

[215] Lahiri, J. and Mukherji, A., Phys.Rev. 153, 386 (1967).

[216] Gupta, A., Roy, H.P. and Mukherjee, A., Int.J.Quantum Chem. 9, 1 (1975).

[217] Reinsch, E. and Meyer, W., Phys.Rev. A 18, 1793 (1978).

[218] London, F., Z.Phys. 63, 245 (1930).

[219] London, F., Z.Phys.Chem. (B)11, 222 (1930).

[220] Slater, J.C. and Kirkwood, J.G., Phys.Rev. 37, 682 (1931).

[221] Margenau, H., Phys.Rev. 38, 747 (1931).

[222] Pauling, L. and Beach, J.Y., Phys. Rev. 47, 686 (1935).

[223] Margenau, H., Phys.Rev. 56, 1000 (1939).

[224] Margenau, H., Rev.Mod.Phys. 11, 1 (1939).

[225] Dalgarno, A. and Kingston, A.E., Proc.Phys.Soc. (London), 78, 607 (1961).

[226] Kingston, A., Phys.Rev. 135, A1018 (1964).

[227] Barker, J.A. and Leonard, P.J., Phys.Letters 13, 127 (1964).

[228] Langhoff, P.W. and Karplus, M., J.Chem.Phys. 52, 1435 (1970).

[229] Langhoff, P.W., J.Chem.Phys. 57, 2604 (1972).

[230] Langhoff, P.W. and Karplus, M. in The Padé Approximants in Theoretical Physics (eds. Baker,G.A. and Gammel, J.L.), Academic Press, N.Y., 1970.

[231] Bell, R.J., National Physical Laboratory Report, MA 56 (1966).

[232] Bell, R.J., Proc.Phys.Soc. 86, 17 (1965).

[233] Arrighini, G.P., Biondi, F. and Guidotti, C., Mol.Phys. 26, 1137 (1973).

[234] Dalgarno, A., Morrison, I.H. and Pengelly, R.M., Int.J.Quantum Chem. 1, 161 (1967).

[235] Chan, Y.M. and Dalgarno, A., Mol.Phys. 9, 349 (1965).

[236] Frenkel,J., Wave Mechanics, Advanced General Theory, Clarendon, Oxford, 1934.

[237] Mc Lachlan, A.D. and Ball, M.A., Rev. Mod. Phys. 34, 844 (1964).

[238] Dalgarno, A. and Victor, G.A., Proc.Roy.Soc. (London) A 291, 291 (1966).

[239] Langhoff, P.W., Epstein, S.T. and Karplus, M., Rev.Mod.Phys. 44, 602 (1972).

[240] Csanak, G., Taylor, H.S. and Yaris, R., Adv.Atom Mol. Phys. 7, 287 (1971).

[241] Linderberg, J. and Öhrn, Y., Propagators in Quantum Chemistry, Academic Press, London, 1973.

[242] Rowe, D.J., Rev.Mod.Phys. 40, 153 (1968).

[243] Longuet-Higgins, H.C., Discuss.Faraday Soc. 40, 7 (1965).

[244] Arrighini, G.P., Biondi, F. and Guidotti, C., J.Chem.Phys. 55, 4090 (1971).

[245] Arrighini, G.P., Biondi, F. and Guidotti, C., Chem.Phys. 2, 85 (1973).

[246] Karplus, M. and Kolker, H.J., J.Chem.Phys. 41, 3955 (1964).

[247] Dalgarno, A. and Victor, G.A., Proc.Phys.Soc. (London) 90, 605 (1967).

[248] Maeder, F. and Kutzelnigg, W., Chem.Phys.Letters 37, 285 (1976).

[249] Meyer, W., Chem.Phys. 17, 27 (1976).

[250] Amos, A.T. and Yoffe, J.A., Theoret.Chim.Acta (Berlin) 42, 247 (1976).

[251] Chan, Y.M. and Dalgarno, A., Proc.Phys.Soc. (London) 86, 777 (1965).

[252] Alexander, M.H., J.Chem.Phys. 52, 3354 (1970).

[253] Tang, K.T. and Karplus, M., Phys.Rev. 171, 70 (1968).

[254] Arrighini, G.P., and Guidotti, C., Mol.Phys. 28, 273 (1974).

[255] Robb, W.D., J.Phys. B 6, 945 (1973).

[256] Kaneko, S., J.Chem.Phys. 56, 3417 (1972).

[257] Langhoff, P.W. and Karplus, M., J.Chem.Phys. 53, 233 (1970).

[258] Briggs, M.P., Murrell, J.N. and Salahub, D.R., Mol.Phys. 22,907 (1971).

[259] Margenau, H., Phys.Rev. 64, 385 (1943).

[260] Lamanna, U.T., Guidotti, C. and Arrighini, G.P., J.Chem.Phys. 67, 604 (1977).

[261] Zeiss, G.D. and Meath, W.J., Mol.Phys. 30, 161 (1975).

[262] Haugh, E.F. and Hirschfelder, J.O., J.Chem.Phys. 23, 1778 (1955).

[263] Langhoff, P.W. and Chan, S.W., Mol.Phys. 25, 345 (1973).

[264] Amos, A.T. and Yoffe, J.A., Chem.Phys.Letters 39, 53 (1976).

[265] Baker, G.A., Jr., Advan. Theoret. Phys., 1, 1 (1965).

[266] Wall, H.S., Analytic Theory of Continued Fractions, Van Nostrand Princeton, N.J., 1948.

[267] Baker, G.A., Jr., in The Padé Approximants in Theoretical Physics (eds. Baker, G.A. and Gammel, J.L.), Academic Press., N.Y., 1970.

[268] Tang,K.T., Phys.Rev. 177, 108 (1969).

[269] Tang,K.T., Phys.Rev. A 1, 1033 (1970).

[270] Salem, L., Mol. Phys., 3, 441 (1960).

[271] Shoat, J.A. and Tamarkin, J.D., The Problem of Moments, American Mathematical Society, Providence, R.I., 1950, Mathematical Surveys, Vol. 1.

[272] Gordon, R.G., J.Chem.Phys. 48, 3929 (1968).

[273] Gordon, R.G., J.Math.Phys. 9, 655 (1968).

[274] Gordon, R.G., J.Math.Phys. 9, 1087 (1968).

[275] Gordon, R.G., Intern.J.Quantum Chem. 2, 151 (1968).

[276] Wheeler, J.C. and Gordon, R.G., J.Chem.Phys. 51, 5566 (1969).

[277] Mc Quarrie, D.A. Tereby, J. and Shire, S.T., J.Chem.Phys. 51, 4688 (1969).

[278] Starkschall, G. and Gordon, R.G., J.Chem.Phys. 54, 663 (1971).

[279] Marquardt, D.W., J.Soc.Ind.Appl.Math. 11, 431 (1963).

[280] Melton, L.A. and Gordon , R.G., J.Chem.Phys. 51, 5449 (1969).

[281] Dantzig, G.B., Linear Programming and Extensions, Princeton U.P., Princeton, N.J., 1963.

[282] Futrelle, R.P. and Mc Quarrie, D.A., Chem.Phys.Letters 2, 233 (1968).

[283] Weinhold, F., J.Phys. A1, 305, 655 (1968).

[284] Weinhold, F., J.Phys. B2, 517 (1969).

[285] Tang, K.T., J.Chem.Phys. 49, 4727 (1968).

[286] Kramer, H.L., J.Chem.Phys. 53, 2783 (1970).

[287] Kramer, H.L. and Herschbach, D.R., J.Chem.Phys. 53, 2792 (1970).

[288] Pack, R.T., Chem.Phys.Letters 49, 3224 (1968).

[289] Goscinski, O. and Brändas, E., Intern.J.Quantum Chem. 5, 131 (1971).

[290] Abdulnur, S.F., J.Chem.Phys. 58, 4835 (1973).

[291] Murrell, J.N. and Shaw, G., J.Chem.Phys. 49, 4731 (1968).

[292] Deal, W.J. and Young, R.H., Intern.J.Quantum Chem. 10, 419 (1976).

[293] Magnansco, V., Figari, G. and Battezzati, M., Mol.Phys. 34, 1201 (1977).

[294] Chalasinski, G. and Jeziorski, B., Mol.Phys. 27, 649 (1974).

[295] Langhoff, P.W., Chem.Phys.Letters 20, 33 (1973).

[296] Jacobi, N. and Csanak, G., Chem.Phys.Letters 30, 367 (1975).

[297] Koide, A., J.Phys. B 9, 3173 (1976).

[298] Battezzati, M. and Magnasco, V., J.Chem.Phys. 66, 3739 (1977).

[299] Mc Weeny, R., Proc.Roy.Soc. A253, 242 (1959).

[300] Bonham, R.A., Peacher, J.L. and Cox, H.L., Jr., J.Chem.Phys. 40, 3083 (1964).

[301] Inokuti, M., Rev.Mod.Phys. 43, 297 (1971).

[302] Lassettre, E.N., J.Chem.Phys. 43, 4479 (1965).

[303] Csanak, G. and Taylor, H.S., Phys.Rev. A6, 1843 (1972).

[304] Lassettre, E.N., J.Chem.Phys. 57, 4357 (1972).

[305] Lamm, G. and Szabo, A., J.Phys. B 10, 1967 (1977).

[306] Au, C.K., J.Phys.B 11, 2781 (1978).

[307] Chalasinski, G. and Jeziorski, B., Mol.Phys. 32, 81 (1976).

[308] Jeziorski, B. and van Hemert, M., Mol.Phys. 31, 713 (1976).

[309] Farrar, J.M. and Lee, Y.T., J.Chem.Phys. 56, 5801 (1972).

[310] Chalasinski, G. and Jeziorski, B., Theoret.Chim.Acta (Berlin) 46, 277 (1977).

[311] Alexander, M.H. and Salem, L., J.Chem.Phys. 44, 2943 (1966).

[312] Certain, P.R., Hirschfelder, J.O., Kolos, W. and Wolniewicz, J.Chem.Phys., 49, 24 (1968).

[313] Bowman, J.D., Ph.D.Thesis, University of Wisconsin, Theoret. Chem. Institute Rep. WIS-TCI-463.

[314] Piela, L., Intern.J.Quantum Chem. 5, 85 (1971).

[315] Jeziorski, B. and Piela, L., Acta Phys. Pol. A42, 177 (1972).

[316] Piela, L. and Jeziorski, B., Acta Phys. Pol. A42, 185 (1972).

[317] Magnasco, V., Figari, G. and Battezzati, M., Chem.Phys.Letters 50, 138 (1977).

[318] Magnasco, V., Figari, G. and Battezzati, M., Mol.Phys. 34, 1201 (1977).

[319] Kreek, H. and Meath, W.J., Mol.Phys. 22, 915 (1971).

[320] Daudey, J.P., Claverie, P. and Malrieu, J.P., Intern.J.Quantum Chem. 8, 1 (1974).

[321] Conway, A. and Murrell, J.N., Mol.Phys. 27, 873 (1974).

[322] Murrell, J.N. and Varandas, A.J.C., Mol.Phys. 30, 223 (1975).

[323] Farrar, J.M., Lee, Y.T., Goldman, V.V. and Klein, M.L., Chem. Phys. Letters 19, 359 (1973).

[324] Stevens, W.J., Wahl, A.C., Gardner, M.A. and Karo, A.M., J.Chem. Phys. 60, 2195 (1974).

[325] Barker, J.A., Fischer, R.A. and Watts, R.O., Mol.Phys. 21, 657 (1971).

[326] Cohen, J.S. and Pack, R.T., J.Chem.Phys., 61, 2372 (1974).

[327] Lloyd, J. and Pugh, D., Chem.Phys.Letters 26, 281 (1974).

[328] Kim, Y.S. and Gordon, R.G., J.Chem.Phys. 61, 1 (1974).

[329] Rae, A.I.M., Chem.Phys.Letters 18, 574 (1973).

[330] Kochanski, E., Chem.Phys.Letters 10, 543 (1971).

[331] Kochanski, E., Chem.Phys.Letters 15, 254 (1972).

[332] Kochanski, E., J.Chem.Phys. 58, 5823 (1973).

[333] Daudey, J.P., InternJ.Quantum Chem. 8, 29 (1974).

[334] Popkie, H., Kistenmacher, H. and Clementi, E., J.Chem.Phys. 59, 1325 (1973).

[335] Matsuoka, O., Clementi, E. and Yoshimine, M., J.Chem.Phys. 64, 1351 (1976).

[336] Kistenmacher, H., Popkie, H., Clementi, E. and Watts, R.O., J. Chem.Phys. 60, 4455 (1974).

[337] Amos, A.T., Chem.Phys.Letters 5, 587 (1970).

[338] Lekkerkerker, H.N.W. and Laidlaw, W.G., J.Chem.Phys. 52, 2953 (1970).

[339] Matsen, F.A. and Junker, B.R., J.Phys.Chem. 75, 1878 (1971).

[340] Chipman, D.M., doctoral dissertation, University of Wisconsin, Theoret. Chem. Institute, Rep. WIS-TCI-459.

[341] Mann, A., Chem.Phys.Letters 32, 363 (1975).

[342] Matsen, F.A., Advan. Quantum Chem. 1, 59 (1964).

[343] Matsen, F.A., J.Am.Chem.Soc. 92, 3525 (1970).

[344] Hirschfelder, J.O. and Silbey, R., J.Chem.Phys. 45, 2188 (1966).

[363] Masad, N., J.Am.Chem.Soc. 92, 3555 (1970).
[364] Hirschfelder, J.O. and Silbey, R., J.Chem.Phys. 45, 2188 (1966).